ADVANCED BIOTECHNOLOGY

EDITORS

Digumarti Bhaskara Rao

Digumarthi Harshitha

K.R.S. Sambasiva Rao

1999

DISCOVERY PUBLISHING HOUSE

NEW DELHI 110 002

Published by:
Tilak Wasan

DISCOVERY PUBLISHING HOUSE PVT. LTD.
4383/4B, Ansari Road, Darya Ganj
New Delhi-110 002 (India)
Phone : +91-11-23279245, 23253475, 43596065
E-mail : discoverypublishinghouse@gmail.com
sales@discoverypublishinggroup.com
web : www.discoverypublishinggroup.com

***First Edition:* 1999**

***Reprinted:* 2020**

ISBN: 978-81-7141-516-8

Advanced Biotechnology

Printed at:
Infinity Imaging Systems
Delhi

FOREWORD

Over the years there has been a gradual transition from the use of traditional recipes and empirical methods to the scientific understanding of the processes that underlie them. The European Federation of Biotechnology has defined biotechnology as "The integrated use of biochemistry, microbiology and engineering sciences in order to achieve technological (industrial) application of the capabilities of micro-organisms, cultured tissue cells, and parts there of."

In short, biotechnology is the application of biological agents in either manufacturing industry or service operation. The use of the term biotechnology may imply a single subject but the essence of biotechnology is its multidisciplinary nature requiring a wide range of science and engineering imputs. The ultimate success of biotechnology is dependent upon advances in and support for the fundamental sciences which form its substratum. The growth of biotechnology in recent years has been so fast and dramatic that the general public understanding of biotechnology and its implications has failed to keep pace with its progress. The capacity to manipulate genetic material in both the laboratory and the field has given us creative powers about the nature of life and its secrets, the more we reduce living organisms to malleable structures open to experimentation and manipulation. Since the application in the engineering of genetic material, the world has witnessed a stunning parade of progress in health care, industrial and agricultural applications, and research into enduring questions in basic biology. Only rarely in the history of science has an area of investigation moved so rapidly from basic research to application as has biotechnology.

In this volume, some features of microbiology, biochemistry, genetics, and engineering which we believe have a significant bearing on the education and development of biotechnology are highlighted.

We hope this book will stimulate the application of scientific and engineering principles to the processing of materials by biological agents in the service of the most urgent human needs.

ZOENEK REHACEK

SUSAN BIGGIN

PREFACE

The field of biotechnology is a multidisciplinary area of applied science which has received input from a variety of disciplines, viz., microbiology, biotechnology, genetics and biochemical engineering. The biotechnology is defined as the integrated use of biochemistry, microbiology and engineering sciences in order to achieve technological application of the capabilities of microorganisms, cultured tissue cells and parts thereof.

In recent years, much research is going on in the field of biotechnology. Many universities have introduced biotechnology at under-graduate and post-graduate levels. To get acquaintance with the latest trends in biotechnology, this book has been brought out.

The authors from different countries have contributed their valuable knowledge for the benefit of researchers and students. The users of this book will understand many a new phenomenon in the field of biotechnology.

We hope this book will generate a great interest in different dimensions of biotechnology.

Dr. D. BHASKARA RAO
Ms. D. HARSHITHA
Dr. K.R.S.S. RAO

ACKNOWLEDGEMENTS

I am thankful to
the director and staff of
UNESCO-ROSTE
Venice, Italy
for giving me permission to
reproduce the articles from
the Technical Report No. 12 entitled
'Modern Biotechnology'
in this book
for the benefit of
biotechnologists, researchers and students.

Ms. D. HARSHITHA
Access

CONTENTS

CONTRIBUTORS

ADLER, I.	Henkel KGaA, D 4000 Dusseldorf 1, Germany.
BELLGARDT, K.H.	Institute for Technical Chemistry, University of Hanover, Germany.
BLAZEJ, A.	Slovak Technical University 81000 Bratislava, Slovakia.
DITL, P.	Czech Technical University, 16628 Prague, Czech Republic.
ETTLER, P.	Institute of Microbiology, Czechoslovak Academy of Sciences, 14220 Prague, Czech Republic.
FIECHTER, A.	Institute for Biotechnology, ETH Zurich, CH 8093 Zurich, Switzerland.
FORT, I.	Czech Technical University, 16628 Prague, Czech Republic.
GRIFFITHS, B.	PHLS-CAMR, Porton Down, Salisbury, England U.K.
HOFFMANN, P.	DSM - German Collection of Micro-organisms and Cell Cultures, D 3300 Braunschweig, Germany.

JAHNKE, K.D.	DSM- German Collection of Micro-organisms and Cell Cultures, D 3300 Braunschweig, Germany.
KOUZMINOV, V.	Chief, UNESCO-ROSTE, Regional Office for Science and Technology for Europe, 1262/A Dorsoduro, 30123 Venice, Italy.
LUBBERT, A.	Institute for Technical Chemistry, University of Hanover, Germany.
MALIK, K.A.	DSM - German Collection of Micro-organisms and Cell Cultures, D 3300 Braunschweig, Germany.
MORRIS, C.B.	ECACC, PHLS-CAMR, Porton Down, Salisbury, England, U.K.
REHACEK, Z.	Institute of Microbiology, Czechoslovak Academy of Sciences, 14220 Prague, Czech Republic.
ROEHR, M.	Institute of Biochemical Technology and Microbiology, Technical University of Vienna, A 1060 Vienna, Austria.
ROUSAR, I.	Czech Technical University, 16607 Prague, Czech Republic.
SCHEPER, T.	Institute for Technical Chemistry, University of Hanover, Germany.
SCHMAUDER, H.P.	Friedrich-Schiller Universitat Jena, D 06900 Jena, Germany.
SCHUGERL, K.	Institute for Technical Chemistry, University of Hanover, Germany.

SCHUMACHER, H.M.	DSM - German Collection of Micro-organisms and Cell Cultures D 3300 Braunschweig, Germany.
SEICHTER, P.	Research Institute of Chemical Equipment 66089 Brno, Czech Republic.
SONNLEITNER, B.	Institute for Biotechnology, ETH Zurich, CH 8093 Zurich, Switzerland.
TORELLA, F.	University of Murica, Spain.
VANECEK, V.	Czech Technical University, 16628 Prague, Czech Republic.
ZIMMERMAN, B.K.	Spectrum International SF 02631 ESPOO Finland and Renvall Institute, Helsinki University, Helsinki, Finalnd.

1

BIOTECHNOLOGY IN THE NINETIES

A. FIECHTER
Institute for Biotechnology, ETH Zurich, Switzerland

It is nearly 30 years since I attended the second symposium on *Continuous Cultivation of Micro-organisms* organized by Academician Ivan Malek, and held in Prague. Three European groups, primarily the Prague group but also the Swedish and the British group from Porton Down, contributed from various fields of microbiology. Cell dynamics and bioregulatory effects were in the foreground. It is not the aim of this paper to discuss the science of yesteryear. However, future is the result of the past, and the innovations of today will coin the future developments.

In the fifties and sixties, the ground was laid for the investigations of the living cell. Chemostat methods became a new approach to uncovering the hidden secrets of life. In parallel, the newly born molecular biology was steadily growing and in 1973 led to the first cloning of a human gene into *E. coli.* Also in the sixties supercomputing started to develop. Progress made in engineering science during the eighties together with the developments of molecular biology spawned present-day biotechnology. The recombinant product market of pharmaceuticals is now coming into existence after a long gestation period of R & D investment without any substantial return of money. Up to Summer 1990, some 123 recombinant products of human genes had been successfully brought into preclinical or clinical testing, 19 of them already on the market (see Bike, Braunschweig, June 1990). This impressive number has been extended by almost 50 additional recombinant products within the last 12 months, alone. They comprise predominantly growth factors, immune modulators, hormones, inhibitors, receptors, vaccines, and blood proteins for actual or potential treatment of humans or animals. As seen in Table 1.1, the market for rec-bioproducts was estimated to grow from US$ 1-2.2 billion in 1985 to US$ 6-10 billion in 1990, with a prediction of US$ 40-64 billion for the year 2000. The corresponding figures for 1990 have actually been reached, confirming the strong increase in value of the rec-bioproducts on the market, mainly products of human genes like tPA, IFN or EPO (erythropoietin). It should, however, be noted that the worldwide demand for quite a few of these recombinant products lies in the range of a few grams to kilograms, and that developments originating from different biotechnological methods can finally lead to the same application showing similar effects. A good example is the case of tPA whose profit predictions had to be corrected downwards after governing the market for years, when scientists showed that streptokinase in combination with aspirin has the same effect for a tenth of the price. Nonetheless, the currently observed fast speed-up in launching new recombinant products, in particular by American and Japanese companies, confirms the optimistic predictions of the economists.

TABLE: 1.1
ESTIMATED MARKET DEVELOPMENT FOR RECOMBINANT BIOPRODUCTS (ANNUAL GROWTH RATES: 23-32%).

Year	*Estimated market for rec-bioproducts (in US$ billion)*
1985	1-2.2
1990	6- 10
2000	40 - 64

There is no doubt that the future spectrum will also be widened to non-medical products, especially enzymes for applications in the food industry and the public sector. Chymosine is a prominent example of such an application.

An important field of biotechnological activity is the degradation of recalcitrant or toxic compounds, aimed at either elimination of the toxic substance in question, or its introduction into a controlled circulation. Two different strategies have been worked out and are currently being tested. The first is based on the design of powerful recombinant strains suitable for degrading the corresponding substance. So far, however, hardly any established processes on this basis have been found. This is due either to problems of genetic instabilities eventually leading to an inactive population that has lost the basis of its degradational capacity, or else to inadequate micro-environmental conditions. Relevant culture parameters such as temperature, pH, or oxygen partial pressure are either not known or cannot be kept constant enough to allow for efficient degradation of the toxic product. To sum up, it can be stated that–seen in this light–the process development itself is lagging behind.

The second strategy is based on the assumption that nature has developed strains of cells able to degrade basically any toxic substance found. Thus is it only a question of a successful enrichment of these particular cells isolated from their natural habitat. However, this has been quite a difficult task to accomplish even when searching in promising sources like sewage sludge or soil samples from dump sites containing the toxic substance. Recently, this enrichment procedure has been greatly simplified by the introduction of nutristat technology, where a constant selection pressure originates from a minimal amount of toxic substance in the culture medium. The concentration of this agent is precisely controlled to avoid killing of the desired cells by overdosage. The lower and upper concentration limits depend on the substance under investigation and can be very close, which requires sophisticated on-line concentration measurements. Initial successful experiments have been carried out on fluorated biphenyles.

Another important biotechnological area of activity is the treatment of sludge in an aerobic thermophilic bioprocess operating at a temperature of 65°C. The process in which residual enterobacteria are at a concentration below 100 per ml and no worm eggs are detected has undergone rapid realization in Switzerland, where 8 plants are in operation and an additional 12 are under construction or at the planning stage.

In the technical field, a breathtaking leap forward has taken place in real-time analytics for physical, chemical, and even biological reaction parameters of variables. Powerful on line analysis systems such as GC (gas chromatography), HPLC (high liquid chromatography), MS (mass spectroscopy), FIA (flow injection analysis) and NMR (nuclear magnetic resonance) are being used. In particular, the FIA system allows the detection of substances of intermediary products with system components which cannot be sterilized. As illustrated in Table 1.2, not only has the signal frequency been increased dramatically, but also the precision of the signal has entered a new dimension with errors below 1% of the dynamic range.

On the other hand, culture containment itself has undergone a series of drastic improvements. An availability of several months is more the rule than the exception. High mass transfer rates and short mixing times (< 1 s) allow for high cell productivity of 20 to 22 g/h.

In parallel, impressive progress is being made on engineering, analytical, and informatic levels. Synthetic membrane research allows integration in batch, repetitive batch, or continuous mode. Efficient cell recycling and on-line product or gas removal are being achieved, and becoming standard in research and production systems.

Finally, sophisticated software and dedicated computer architecture allow not only high precision process control, but provide full automation potential for production or even extended research programmes.

An example of such a programme is the on-line development of an optimal medium composition, depicted in Fig. 1.1. Each component of the medium, such as carbon sources, vitamins, or trace elements is stored as pure solution and pumped into the reactor according to the actual control strategy. This allows the cell metabolism to be influenced, for example, by switching carbon sources or setting up object oriented limitations.

A first control loop for physical and chemical variables is effected by a slave computer which, by delivering relevant data to a supervisory computer, allows the recognition of patterns of parameters leading to a control loop for the biological variables. The on-line medium strategy is only limited by the fact that substances can only be added and not withdrawn.

Further, as the technique is based on the flow of substances into the reactor, it can only be propagated in fed batch or continuous culture.

TABLE: 1.2.
ON-LINE VARIABLES AND PARAMETERS IN BIOTECHNOLOGICAL PROCESES (SONNLEITNER, 1991).

Variable	*Type/ range*	*Units*	*Accuracy / Precision*
Physical Variables			
temperature	0-150	°C	0.01%
rpm	0-3000	min^{-1}	0.2%
weight	90-100	kg	0.1%
	0-1	kg	0.01%
liquid flux	0-8	m^3h^{-1}	1%
	0-2	kg h^{-1}	0.5%
dilution rate	0-1	h^{-1}	< 0.5%
gas flux	0-2	vvm	0.1%
foam	on/off	-	-
bubbles	on/off	-	-
level	on/off	-	-
Chemical Variables			
pH	2-12	units	0-1%
pO2	0-100	% sat	1%
pCo2	0-100	mbar	1%
exhaust O2	16-21	%	1%
exhaust CO2	0-5	%	1%
fluorescence	0-5	V	-
redox	-.6-.3	V	0.2%
MS: volatiles			membrane
methonaol, ethanol	0-10		
acetone	0-10	g l^{-1}	1-5%
butanols	0-10		
on line FIA:			=f(dilution)
glucose (< 100 h^{-1})	0-100		< 2%
NH_4^+ (< 20 h^{-1})	0-10	g l^{-1}	1%
PO4^{+++}(< 15 h^{-1})	0-10		1-4%
GLU (< 20 h^{-1})	0-10		< 2%
on line HPLC:			
phenols (< 5 h^{-1})	0-100	mg l^{-1}	2-5%
phthalates	0-100	mg l^{-1}	2-5%
organic acids	0-1	g l^{-1}	1-4%
erythromycins	0-20	g l^{-1}	< 8%
by products	0-5	g l^{-1}	2-5%
on line GC:			capillary
acetic acid (< 10 h^{-1})	0-5		2-7%
acetoin	0-10		< 2-5%
butanediol (R.S)	0-10		< 8%

{cont.}...

meso-butanediol	0-10	g l^{-1}	< 8%
ethanol	0-5		< 2%
glycerol	0-1		< 9%
Biological Variables			
RQ	0-5-20	M M^{-1}	error propagation
x: OD-sensors	0-10-0	AU	highly variable
physiological state	software	-	reference patterns
μ (=D)	0-1	h^{-1}	< 0-5%

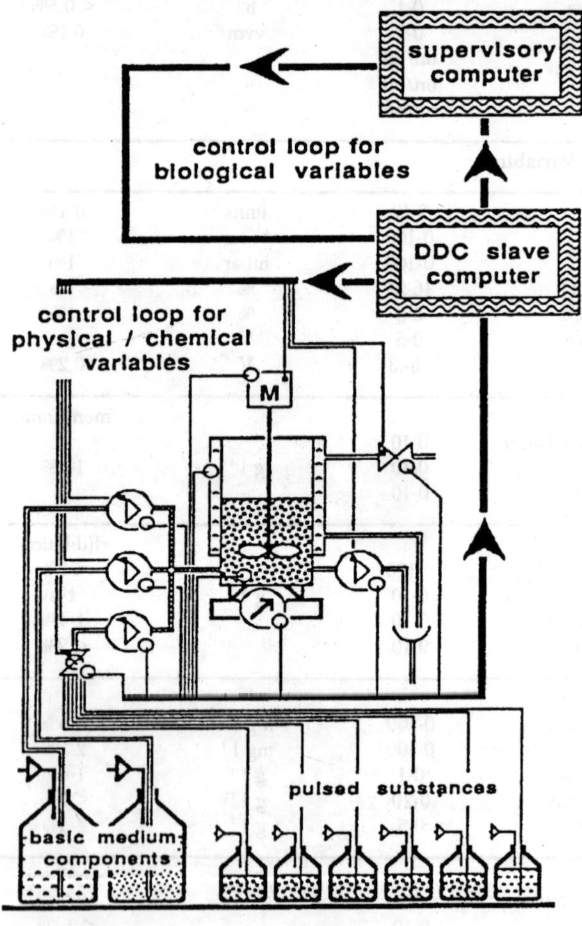

Figure 1.1. On line medium composition (Locher et al., Biotechnol, submitted).

The most recent results released from such high-teach cultivation systems demonstrate new possibilities for fundamental research in biology, such as studies of the cell cycle or metabolic control.

Examples range from the determination of repetitive metabolic fluxes in synchronized yeast cultures depending on the cell cycle status, to the generation of new control parameters for eukaryotic cell cultures like CHO.

Experimental access is now provided to the undisturbed and fully active cell by these non invasive concepts. Consequences of these new methodologies for the nineties are substantial, as illustrated in Table 1.3.

Trends for concerted efforts in the development of overproducing host cell lines cannot be overlooked. In 1990, only some 10 to 15% of the biosynthetic capacity of the cell were exploited in the production plants. The design and realization of expression systems for recombinant protein production is a time consuming process.

Following this line of argument, one could therefore expect development of the 'top performance host' (TPH) with a maximum synthetic efficiency, maximum ATP generation, and maximum respiration. The cell line would achieve an optimal partition of energy for growth versus product synthesis.

Fast structure resolution of proteins as a prerequisite for protein modelling and fast genetic adjustments to meet the requirements of the bioprocess are the basis for TPH and call for the complete characterization of its genome.

The efforts of several megagenome sequencing projects initiated several years ago pointed in this direction–such as the sequencing of the complete yeast genome, or the human genome sequencing project headed by HUGO. It is encouraging to see that genes like that responsible for the Fragile X syndrome, leading to mental retardation, could be isolated and characterized well before the expected time.

ACKNOWLEDGMENTS

The assistance of Wolfgang Seghezzi in the preparation of the manuscript has been highly appreciated.

TABLE: 1.3. SOME (RECENT) HIGHLIGHTS IN BIOTECHNOLOGY

Topic	*Comments*
Cloning of higher plants	Crop improvement: * Efficiency of production * Adaptation to environmental parameters * Replacement of pesticides by transgenic resistances
Heterologous animals	Improvement of animal productivity Improvement of fertility, robustness heterologous expression of foreign (glycosylated) proteins
Glycobiology	Most proteins for human/animal treatment: Antibodies] Carbohydrate Peptide hormones] chemistry ..tPA] will become ..Receptors] more ..Immuno-modulators] important
Fusion proteins	Promoter of bone cell growth GM-CSF + IL-3 (Granulocyte macrophage colony stimulating factor fused to Interleukin-3) Cancer, AIDS, rheumatic arthritis Diphtheria toxin + IL-2
Antisense technology	Inhibitors Antagonists
Ribozymes	RNA with 'enzymatic' activity (Cech and Altman, 1990, Nobel Prize) potential: deliberate splitting of mRNA by RNA (HIV, herpes, hepatitis B,....)
Gene therapy	1990: first treatment of genetic disease by cloning of adeno-sinedeaminase into deficient leucocytes 1991: cystic fibrosis
Biocare	Combination of biotechnology with medical care, mostly with (recombinant) cells (of the patient); e.g., gene therapy in the frame of a tailored medical treatment concept

2

RAW MATERIAL STRATEGIES

M. ROEHR
Institute of Biochemical Technology and Microbiology, Technical University of Vienna, Austria

An inspection of the main products of biotechnology shows that a considerable proportion of traditional as well as modern products are high-volume/low-price (HVLP) products (Fig. 2.1). These are characterized by the fact that raw material costs may amount to 50...80% of total costs.

Selling prices of HVLP products easily come under pressure, especially if they have to compete with synthetic products, as is the case with alcohols or similar commodities, and some amino acids. In most cases it is the price of the raw material that determines the degree of competitiveness.

Nevertheless, problems of provision and preparation of raw materials have to be considered a more or less neglected area compared with other complex activities of biotechnologists. In the biotechnological literature, therefore, very few examples can be found that treat this subject adequately.

The reason may be seen in the fact that most of the raw materials in biotechnology are products, by-products, or even waste materials of agriculture and the food industry or forestry, and are therefore considered to lie beyond the scope of the average biotechnologist.

TABLE: 2.1
CONTRIBUTION OF SUBSTRATE COSTS IN FERMENTATION PROCESSES

Product	Raw materials	Substrate costs (% total prod. costs)
Single cell protein	Methanol	63
Single cell protein	Molasses	62
Ethanol	Sugar cane	42
Ethanol	Corn	54
Ethanol	Crop residue	52
Acetone+butanol	Molasses	33
Itaconic and	Molasses	49
Citric acid	Molasses	33
Cellulase enzymes	Lignocellulose	51
Bacitracin	Sugar	25

Source: Data, 1987

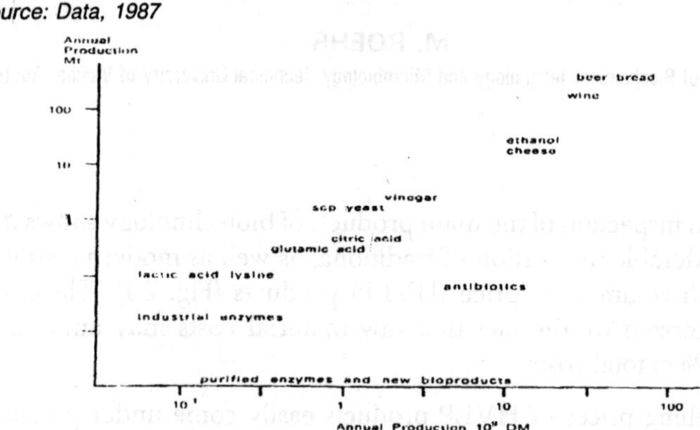

Fig. 2.1: Industrial production. Source: Hepner and Male, 1982.

A more complete knowledge, however, of raw material strategies has become increasingly important for several reasons, of which the main three are mentioned and discussed in the following:

(i) changes in availability, quality, and prices of the traditional raw materials of fermentation industries
(ii) problems of environmental protection, global impacts of consumption of resources and energy
(iii) overproduction in agriculture and the food industries of the developed countries.

Biotechnologists today have to take these items into consideration when designing strategies to develop or run new processes or scale up old ones. Several examples serve to illustrate the meaning of these three issues.

For a long time, typical raw materials in European fermentation industries were byproducts of sugar manufacture, i.e., molasses (including hydrol molasses from starch hydrolysis). In the European Community, molasses is produced in an amount of four million tonnes, more than two million tonnes being imported. Of these six million tonnes, 3.5 million tonnes are used as animal feed, 2 million tonnes for the production of yeast and ethanol, and 0.5 million tonnes for citric acid manufacture. Although world sugar industries are producing increasing amounts of beet and cane sugar, the amount and especially the quality of molasses have decreased considerably. This is the result of improvements in sugar manufacturing technology yielding molasses with lower sugar contents and concomitantly high proportions of non-sugar compounds. Such molasses are no longer suitable for certain fermentation processes, e.g., baker's yeast manufacture or citric acid fermentation.

The alternative strategy, now, is to integrate sugar manufacture with fermentation by producing and utilizing syrups from the first centrifugation stages–so-called green-syrups. These can be used with little adaptation in various fields and apparently can be supplied in greater amounts.

In the evaluation of molasses as substrates, one should always observe the price differential between molasses and sucrose, since the costs of disposal of the non-sugar portion of molasses may be significant, nowadays, for the preference of 'free', i.e., surplus, sucrose– if available in sufficient amounts on the market.

TABLE: 2.2
COMPOSITIONS OF DIFFERENT MOLASSES

	Beet (%)	*Beet low quality* (%)	*Green syrup* (%)
Dry weight	78.0-85.0	76.8-86.8	77.7-84.0
Total carbohyd.	48.0-58.0	44.8-51.2	53.0-68.3
Non-sugars	20.0-37.0	25.6-42.0	9.4-31.0
Nitrogen	0.2-2.8	3.2-3.6	var.
Ash	4.0-8.0	var.	var.

Source: Austrian Sugar Research Institute, Kominek.

Whey as another potential raw material presents different problems. The world's dairy industry produces increasing amounts of whey–around 10^8 tonnes per year –of which up to one half is used as such in animal feeding, 25-50% are processed to whey concentrates and to lactose with subsequent disposal of the residues, and about 25% are disposed of as such. One tonne of whey causes environmental pollution equivalent to 400 people! It should be noted, however, that whey is extremely perishable.

Thus, collecting of larger amounts from larger areas is hardly feasible. Here a new technology has been established within the last two decades, using several novel techniques directed at concentrating those components of whey that are potentially valuable, but low in concentration and separation methods. These techniques are based on size or electric charge of the constituents of whey, and have been developed yielding whey-protein concentrates and individual compounds of high quality on an industrial scale.

An interesting and promising technology, invented by a French group led by Maubois (1988), merits particular mention. This method enables the production of high purity whey protein concentrates and that of individual whey proteins. It was found that the main obstacle in purifying whey constituents is the content of residual fat, which occurs mainly as phospholipoproteins. These can be separated by precipitation as a calcium ion induced aggregate upon heat treatment. The resulting precipitate can be removed by microfiltration and sold, for example, as an emulsifier. The defatted whey is much more susceptible to ultrafiltration, since fouling of the respective membranes is greatly reduced. Whey protein concentrates derived thereof with 100 g protein per kg of product are excellent food components, but 90 to 95% (N* 6.38/dm) concentrates are also obtainable, and it is possible to produce, for example, pure alpha-lactalbumine and beta-lactoglobuline. The by-products of such value-adding steps–and this is good strategy–are cheap and abundant fermentation substrates for, for example, ethanol (even wine) or microbial biomass production on an industrial scale.

The second type of traditional carbohydrate raw material is starch from various starch-containing crops. Several novel strategies have been developed, mainly during the seventies and eighties, in the endeavour to produce ethanol, so-called bio-ethanol, from various raw materials, as an alternative source of fuel and chemical feedstock. The need for improvements in all stages of the process of conversion of carbohydrates to fuel ethanol stems from the fact that there is a great controversy with respect to the energy as well as the cost-price balance of this process in comparison to the use of fossil fuels.

The main strategies are as follows:

1. Increase the per hectare yield of plant biomass or its starch content by genetic means aiming at creating almost novel plants. It should be kept in mind that for centuries, crops were empirically optimized for their special uses mainly in the food industry. For example, grain was optimized either to yield special flour qualities or to be specially suitable as animal feedstock. It should further be mentioned that arable land on our planet (about 10^{10} hectare) is mainly–75%–used for grain production; nearly 50% of the world's per capita calorie and protein intake are obtained from cereal grains for direct human consumption (FAO, Rome, 1980). This shows, on the other hand, that less has been done regarding, for example, root starch containing plants. Based on crop area to produce a certain product (e.g., alcohol), however, such plants may occupy a rather favourable position.

TABLE: 2.3
CROP AREA FOR A 100,000 T/A ETHANOL PLANT

	1000 hectare
Wheat	214
Corn	136
Rice	125
Sweet potatoes	115
Potatoes	86
Yams	75
Cassava	70
Sugar cane	37

Source: Dimmling, 1979.

2. Improve the traditional technique of starch hydrolysis and also that of alcoholic fermentation, especially aiming at decreasing the energy requirements of the processes. By consequent investigations of all the stages of the processing of starch-bearing material, it was possible to achieve multiple improvements. These include, for example, in the so-called Supramyl Process, mechanical treatment (wet milling) of the starch slurry during the first enzymic liquefying stage, enabling the temperature to be held far below the usual high temperature in the converter without

giving rise to elevated viscosities. This, on the other hand, enables concentrated mashes to be obtained with high ratios of slops recycle. Provided that raw material mixtures with 55-60% dry matter are introduced into the process (e.g., moist, freshly harvested corn or 60% grains plus 40% potatoes), almost complete recycling of slops is possible. This means that instead of evaporating 9-10 litres of water per litre of ethanol, only one litre of water has to be removed. Altogether, the energy balance of ethanol production may be improved by a factor of 2 to 4.

In this context, one of the main improvements of starch conversion technology should be mentioned: the development of enzymes that enable almost complete conversion of starch to crystallizable dextrose. The introduction and production of high-temperature resistant liquefying alpha-amylases and highly active dextrinating and saccharifying enzymes has been a milestone in enzyme technology. This will be mentioned especially in the discussion of a significant problem for decision makers in economic process design. The economic production of the enzymes mentioned above has been the domain of only a few companies, of which some are in small countries such as the Netherlands or Denmark. If one considers that the costs of special enzymes for such conversion processes (but also for numerous others) contribute only very little to the total production costs of the product, it is worthwhile deciding not to invest in reinvestigating these developments but rather to concentrate on the characteristics of the indigenous raw material in order to optimize the process.

TABLE: 2.4
ENZYMES USED IN CONVERSION

Product	*Enzyme protein used (ppm of product)*	*Enzyme costs (% product value)*
Detergents	150 - 200	1
Dextrose	150 - 200	1
Isomerose syrup	150 - 200	3
Ethanol	300 - 400	3
Cheese	003 - 006	0.1- 0.3

Source: Austrup, 1984.

Similar strategies have been pursued in the utilization of lignocellulose materials. Lignocellulose offers one of the most abundant and challenging raw materials. Let us consider some data:

the worldwide annual production of biomass amounts to about 10^{11} tonne, the main portion being lignocellulose, followed by starch (about 10^9 tonne from grain) and sugar (about 10^8 tonne). Only a small proportion of the lignocellulose is utilized (about 2%). As may be seen, however, biomass may well be recognized as a potent chemical feedstock.

TABLE: 2.5

ROLE OF BIOMASS AS A CHEMICAL FEEDSTOCK

World production of:	*Mtonnes*	*Oil equiv. (OE) Mtonnes*	*Into chem. ind. Mtonnes*	*(%)*
Crude oil	3000	3000	135	4.5
Coal	2800	1900	26	1.4
Natural gas	1000	1100	60	5.4
Biomass:				
Wood	1700	600		
Starch equ.	900	300		
Fatty oils	60	20		
	2660	920	19	2.0

Source: Faust, 1985.

There has been much speculation as to whether, or to what extent, lignocellulosic materials could be used as substitute for the primary energy carriers of to-day. Notwithstanding the fact that coal, followed by oil shale and tar sands (besides atomic power) will be able to cover the energy demands of the industrialized countries, if oil supply reaches a critical phase, there is space for lignocellulosic biomass as an alternative fuel (Fig. 2.2).

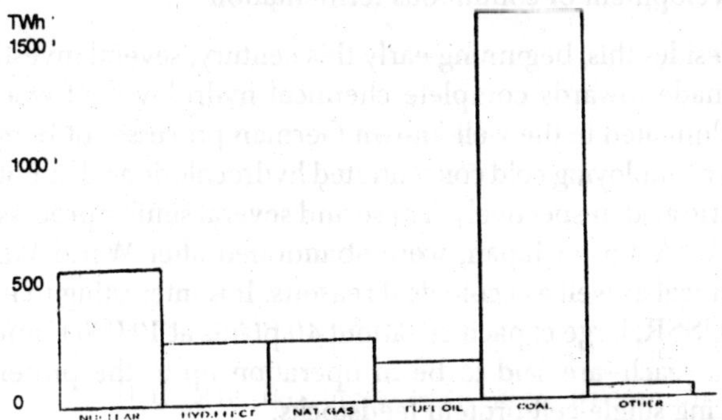

Fig. 2.2: Total US electricity generation (1989). Source: Nat. Geogr. Soc., 1991.

An increasingly important idea behind utilizing renewable resources is the growing disproportion between the emission of carbon dioxide through the combustion of fossil fuels and the ability of plants to recycle carbon dioxide in photosynthesis. Two hundred tonnes of fossil fuels per second are burnt worldwide, producing five hundred tonnes of carbon dioxide!

Apart from this well-known discussion, there are numerous projects to breed and grow so called 'energy-plants' as a source of combustible fuel. The aims are to elaborate strategies to produce fast-growing plants in short rotation forestry, and promising results have been achieved. In particular, the so-called C4-plants should also be mentioned here. Besides this combustion of lignocellulosic waste materials is another target of energy technology.

Lignocellulose, however, could also offer probably one of the cheapest raw materials for the fermentation industries. Being mostly used as timber and fuel (about 85%), only 15% are used in the pulping and cellulose industries. In both areas, waste materials are recovered in large amounts and could be disposed of by being used as a fermentation raw material. Numerous attempts have been made towards this goal. The first lignocellulose-derived feedstock obviously was sulphite waste liquor. It was mainly used for the production of technical-grade ethanol and single-cell protein–depending on the kind of wood and the specific local demands. Again for reasons of environmental protection, the cellulosic industries had to change their technologies to almost completely closed-cycle operations, thereby making sulphite-liquor processing obsolete. Historically, I believe, the fermentative processing of this material was quite instrumental in the development of continuous fermentation.

Besides this, beginning early this century, several investigations were made towards complete chemical hydroloysis of wood. This first culminated in the well-known German processes of Bergius and Scholler, employing cold concentrated hydrochloric acid or hot diluted sulphuric acid, respectively. These and several similar processes, e.g., in the USA and in Japan, were abandoned after World War II for economical as well as ecological reasons. It is interesting to note that in the USSR, large capacities–about 40 plants at 10,000 tonnes sugar per year each–are said to be in operation up to the present time, producing single-cell-protein feedstocks.

The main drawbacks of these processes are their low sugar yields due to various degradation reactions and their high amount of environmental pollution. However, during the last twenty years, numerous research projects have been launched which aim to develop methods of physical pretreatment of lignocellulosic and cellulosic materials in order to facilitate subsequent hydrolysis. Many such methods have been invented during the last two decades. These, in turn, were most valuable for parallel research on the enzymatic hydrolysis of cellulose and delignification of lignocellulose.

Over the last three decades, the potential of cellulose hydrolyzing enzymes has been investigated extensively, and this has lead to the development–although still in the pilot plant stage–of several processes of enzyme-catalyzed cellulose hydrolysis processes, adopted to various raw materials. Examples of such materials are straw, newspaper and similar paper wastes, assorted rubbish, agricultural wastes as well as residues of wood processing. Availability of selected raw materials may be seen from compilations, for example, for Austria or the USA.

TABLE: 2.6
PRETREATMENTS AND SUGAR YIELDS OF LIGNOCELLULOSIC BIOMASS

Method	*Material*	*Maximum sugar yield in 24 h hydrolysis (g/g dm)*
Sodium hydroxide	Wheat straw	0.29
Sulphuric acid	Ryegrass straw	0.30
Ozone	Cotton straw	0.31
Ball milling	Wheat straw	0.26
Compression milling	Poplar	0.50
Steam explosion	Aspen	0.44
Ammonia freeze explosion	Aspen	0.45

Source: Dale, 1987

In Austria, a pilot plant went into operation several years ago. The plant uses steam explosion as pretreatment of various waste materials. Cellulose production has been optimized, especially with respect to the raw material used.

TABLE: 2.7
AVAILABILITY OF WASTE MATERIALS

	Mtonne dw per year *USA*	*Austria*	
Urban wastes	125	1.0	Urban wastes
		1.82	Sawmill wastes
Wood waste	51.8	0.46	Bark
Agricultural residues	430.0	3.24	Agricultural residues
Waste paper	27.3		
Total	634	6.5	Total

Source: Hofreither *et. al.*, 1987.

TABLE: 2.8
GLUCOSE YIELDS IN PILOT PLANT

	g/100 g
Wheat straw	34
Sweet sorghum bagasse	34
Palm press fibres	15
Palm press stalks	24
Potato pulp	54
Beech wood	35
Poplar wood	30
Newspaper	36

Source: Esterbauer.

The dextrose yields are in the range 60-100% of the theoretical value. Raw materials with higher amounts of hemicelluloses may be processed to yield furfural as a valuable by product. In 1988, a sugar cost price of 3.12 Austrian Schillings (about US$ 0.25) per kilogram was given, which is not paying. Efforts were made to reduce enzyme costs since costs of enzyme production, in contrast to other enzymes, are relatively high.

Efforts have also been made to exploit the potential of hemicellulose and lignin. Hemicellulose can be hydrolyzed rather easily to yield mainly pentoses, i.e., xylose. Since xylose is just scarcely metabolized by micro-organisms, the possibility of isomerizing xylose to xylulose has been explored. Xylulose may be fermented to ethanol by a number of yeasts, but productivities as well as yields are low. Economic utilization of lignin on a larger scale must still be considered a challenge to chemists and biotechnologists.

Altogether, biotechnological lignocellulose utilization is still a problem of economics, although optimistic cost analyses exist.

In order to evaluate the economy of raw material production in relation to possible products, it is advisable to determine the proportionate cost price (PCP) of the raw material in the product:

$$\frac{\text{ex-factory price of product-production (= conversion) costs}}{\text{amount of raw material per unit of product}}$$

The following tables show that, for example, sugar produced for citric acid manufacture yields a PCP of 5000 AS per tonne, whereas for single cell protein only 730 AS can be achieved, which is well below market price.

Such prices, of course, are subject to changes depending on quality or purity requirements in manufacturing a certain product.

Several factors contribute to the aim of minimizing costs of raw materials from various sources, especially from lignocellulosics. According to Dale (1987), these may be specified as follows:

- separate as many products as possible from a given material
- separate preferentially the most abundant components
- conduct the process so as to retain the most valuable component
- design the plant to process a high number of different materials
- estimate whether the ex-factory (cost) price of the product(s) approaches at least the two-fold (preferentially three-fold) price of the raw material.

We have learned thus that most economic solutions may be achieved by aiming at multiple products from multiple raw materials applying manifold production steps–in other words, the basic concept tends to the formation of integrated production systems with maximum flexibility as outlined, for example, by the Danish group of Munck (1987), or expressed by an expert group at the Forum Alpbach in the Tyrol in 1988.

All these activities are directed towards a change from strategies based on the abundancy and utilization of cheap energy resources, to a new approach of carefully observing the requirements of environmental protection, decentralized production in highly interdependent systems with a high degree of self-sufficiency with respect to energy, fertilizers, feeding materials, etc.

TABLE: 2.9
PCP OF SUGAR IN VARIOUS PRODUCTS

Product	*Ex-factory price of product (AS/t)*	*Production (conversion) costs (AS/t)*	*PCP (AS/t)*
Ethanol	4500	2500	1140
Lactic acid	18000	12250	5080
Citric acid	20000	13753	5000
SCP	4800	3350	730
Furfural	14000	7230	3080

Source: Hofreither *et. al.*, 1987.

TABLE: 2.10
PRODUCTION COSTS

	Starch glucose AS/kg	*Straw Glucose AS/kg*	*Straw C_5C_6 AS/kg*
Chemicals	0.115	0.172	90 t/d
Energy	0.060	1.723	130 t/d
Personnel	0.150	0.383	= 220 t/d
Fixed costs	0.065	0.327	3.537*130/220=2.09
Capital costs	0.090	0.932	
Total	0.480	3.537	2.09

	Pentose→Furfural→furfuryl-alcohol AS/kg	*AS/kg*	*Pentose→SCP AS/kg*
Chemicals	0.200	3.200	0.833
Energy	3.900	4.415	0.657
Personnel	0.230	0.345	0.115
Fixed costs	0.636	0.805	0.380
Capital costs	2.254	2.855	1.350
Total	7.220	11.620	3.335

	Ethanol AS/kg	*Citric acid H_2O AS/kg*	*Lactic acid AS/kg*
Chemicals	0.042	4.412	1.900
Energy	1.580	5.036	2.560
Personnel	0.115	0.330	0.500
Fixed costs	0.113	0.875	0.890
Capital costs	0.650	3.100	6.400
Total	2.500	13.753	12.250

Source: Hofreither, *et. al.*, 1987.

Munck (1987) has given a thorough account of the drawbacks of present agricultural systems in developed countries, now possibly being replaced by such new ideas as 'utilization of the whole plant' to produce 'multi-purpose products' in decentralized interwoven systems, i.e., the socalled 'processing stations' or 'agricultural refineries' working in closed-cycle operation. An important feature of such systems is the combination of food production and manufacture of various kinds of commodities in adjacent small production units. Traditional examples are the old combinations of alcohol production from grains or potatoes with recycling of slops for animal feeding and utilization of animal off-products as fertilizers on one and the same farm area, as part of an official contract as was found, for example, in Prussia a hundred years ago. More recent examples are the exploitation of sorghum plants for pulp production, as investigated thirty years ago.

While this is a glimpse somewhere into the future, the present situation is characterized by a certain abundancy of agricultural resources in developed countries due to over production.

TABLE: 2.11
COMPOSITION OF SORGHUM PLANTS EXAMINED FOR PULP PRODUCTION

Type of sorghum	*Cellulose (%)*	*Pentosan (%)*	*Lignin (%)*
Grain sorghums	29	24	16
Broomcorns	39	24	15
Sweet sorghums	26	17	10
Sudangrass	43	45	16
Johnsongrass	42	27	17

Source: Lengyel and Anus, 1960.

In 1990, the developed countries made up 24% of the world economy, and were responsible for 50% of the world grain production.

This also means that owing to certain misconceptions in the policy of agricultural and food production, overproduction is continuously increasing, bringing with it an excess supply of agricultural products, mainly for the food industries, and therefore at relatively high cost prices, and these are usable only to a limited extent as feedstocks for the fermentation industries.

In most countries, this surplus has to be exported, and heavily subsidized by the government and farmers themselves. In (the former

FR) Germany, the requirement for subsidizing different agricultural raw materials lies between 800 and 2500 DM per hectare.

In Austria, as another example, the agriculturally utilized area is around 1.5 million hectares yielding a surplus grain production of up to one million tonnes, corresponding to an area of 250,000 hectares.

The respective grain export has to be subsidized to the tune of about four billion Austrian Schilings. If this area were used to produce oil seeds, e.g., rape seeds, as currently being experimented in Austria, the respective amount of oil–t/ha + 2 t of press residues–could be transformed by alcoholysis, for example, to the corresponding rape seed oil methylester. This is an excellent substitute for diesel fuel and could be sufficient substitute for diesel fuel and could be-sufficient to being about self sufficiency in this part. The resulting press residue (fooder) as well as glycerol (0.1 t/ha) as the product of methanolysis, are valuable by products. Besides this, there are plans to produce alcohol from larger amounts of remaining surplus grain and to use this amount for admixing to gasoline instead of exporting.

TABLE: 2.12
PRODUCTION AND CONSUMPTION, 1974-1984, FOR VARIOUS AGRICULTURAL RAW MATERIALS

	Change in Production 1974-1984 (%)	*Change in consumption 1974-1984 (%)*	*Self-sufficiency (%)*
Grain	+24	-3	130
Sugar	+53	-4	147
Milk	+19	+4	115
Wine	+29	+5	120

Source: Willer, 1985

This, on the other hand, may contribute to environmental protection, since the admixed ethanol (5-10%) renders motor-fuel leading unnecessary.

Again, the concomitant yield of animal feed from slops contributes to improved self-sufficiency in feed supply.

Medium-term strategies thus comprise alternative crops and alternative uses of materials. Both alternatives, of course, offer new possibilities for the production of fermentation raw materials at lower prices as components or by-products of production systems.

Within this line of development, starch containing materials will, at least in the medium-term, be the substrates of choice, either as surplus products of the present production policy or as by products of the proposed smaller integrated production units as just mentioned. This may represent an opportunity for less developed countries–to produce or rather to process such cheap raw materials as by-products of the less advanced food production.

The story of alkanes as substrates for microbial processes offers the opportunity to discuss another item.

The possibility of utilizing n-alkanes as substrates for large-scale propagation of micro-organisms initiated a worldwide boom in producing cheap SCP. This only ceased after drastic oil price increases in the mid-seventies and owing to a lack of acceptance by the public in the producer countries. In this context, the expression 'unconventional' substrates and products was coined. Research on the utilization of such unconventional substrates, however, revealed several novel metabolic pathways leading to innovative ways of product formation e.g., organic acids. On the other hand, investigations into the use of methane as a growth substrate led to the utilization of methanol as an especially cheap raw material. The advantages of methanol are its accessibility from a variety of feedstocks (fossil and renewable)–in terms of available carbon, it is the cheapest raw material, the price being around one third that of ethanol for example. A clear disadvantage of methanol is its inhibitory action on micro-organisms at elevated concentrations. Thus only fed batch operations may be applied. Although several possibilities of product formations have been described, so far it is only used for SCP production.

In this context, a more recent concept should be mentioned, namely, the application of multiple feedstocks in fermentation media. The use of multiple substrates, as they occur in nature, has been common experience for a long time. The novel ideas behind modern use of multiple substrates in fermentations is based on a deeper understanding of cellular metabolism. It is known that under conditions of nutrient sufficiency, the response of a micro-organism to multiple substrate supply is diauxic in nature. Under conditions of nutrient deficiency, multiple substrates are frequently utilized simultaneously.

In many cases, different substrates are metabolized by different pathways, thus one substrate may be utilized preferentially as an energy source while the other may serve for anabolic reactions. This can be utilized with the aim of better substrate utilization or with that of inducing reactions leading to special product formations, and this could be a new strategy to produce certain LVHP products.

LVHP products, of course, are much less sensitive to prices of raw materials. Fine chemicals, in most of the cases produced for pharmaceutical purposes, first of all require the application of pure and reproducible raw materials. The amount of raw material applied is limited owing to the relatively small amount of product made. There are certain exceptions, however, and only these will be mentioned.

One of the fields of modern biotechnology is animal cell culture. Here, rather complex production media are applied, and there is a clear tendency towards synthetic media. Several medium constituents can be supplied by fermentation technology, e.g., amino acids, vitamins, etc., but it is also possible to produce certain proteins (serum proteins) by recombinant techniques.

In summarizing, we may conclude: in view of various constraints–environmental protection, agricultural overproduction, need for diversification, progress in technology–demands for cheap and suitable raw materials for the fermentation industries may be satisfied adequately if the necessary socio-economic prerequisites are realized. The biotechnologists' role can be to contribute its scientific base in the search for optimum solutions.

REFERENCES

Arbeitsgemeinschaft Nachwachsende Rohstoffe bei der Verbindungsstelle Landwirtschaft-Industrie eV, *Nachwachsende Rohstoffe*, J. Kordt, Bochum, Germany, 1986.

Aunstrup, K. 1984. Enzyme technology, problems and possibilities. In *Proc. Third Europ. Congr. Biotechnol.* Munich 1984. Chemie, Weinheim. Vol. IV, pp. 143-56.

Barr, T.N. 1981. The world food situation and global grain prospects. Science, 214, 1087-95.

Benet, G. and Rouchouse, J. 1988. Synthetic media in industrial biotechnologies–dream or reality. *Proc. 8th Int. Biotechnology Symposium.* Paris, 1988. Volz. 1, pp. 228-58.

Bundesministerium fur Forschung und Technologie (ed). 1986. *Nachwachsende Rohstoffe.* Bonn, Germany.

Dale, B.E. 1987. Lignocellulose conversion and the future of fermentation biotechnology. *TIBTECH,* 5, 287-91.

Dimmling, W. 1979. Substrates in fermentation technology–situation and economy. In *Microbiology Applied to Biotechnology. Proc. 12th Int. Congr. Microbiol.* Munich, 1978. Chemie, Weinheim. pp. 1-26.

Faust, U. 1985. Bedeutung der Biotechnologie fur Chemie und Energie. In *15th Int. Symp. Zukunftschance Biotechnologie.* Mauterndorf, 1985. Osterr. Ges. land- and Forstwirtschaftspolitik. pp. 121-34.

Hofreither, M.F. Schneider, F. and Weiss C. 1987. Biosprit in Osterreich–eine volkswirtschaftliche Analyse. Universitat Linz.

Lengyel, P. and Annus, E. 1960. Domestic Hungarian plants suitable for pulp manufacture. Papiripar, 4, 100-05-5.

Maubois, J.L. 1988. Whey, its biotechnological signification. *Proc. 8th Biotechnolog. Symp.* Paris, 1988. Vol. II, pp. 814-24.

Munck, L. 1987. A new agricultural system for Europe. *TIBTECH*, 5, 1-4.

Rask, N. 1980. Biomass, its utilization as food and/or fuel, *Proc. IVth Int. Symp. Alcohol Fuels Technol.* Guaruja, 1980. Vol. III, pp. 953-64.

Ratledge, C. 1977. Fermentation substrates. *Proc. Ann. Rep. Ferment.* Vol. III, pp. 49-71.

Rexen, F. and Munck, L. 1984. *Cereal Crops for Industrial Use in Europe.* Report prepared for The Commission of the European Community, EEC and Carlsherg Research Lab., Copenbagen, Denmark.

Schmidt, A. 1982. Biomasse als Rohstoff und Energietrager. *Ost. Chemie Z.*, 83, 79-87.

Shacter, E. 1989. Serum-free media for bulk culture of hybridoma cells and the preparation of monoclonal antibodies. *TIBTEC*, 7, 248-53.

UNIDO Secretariat. 1983. Implications of Biomass Energy Technology for Developing Countries. 7 March 1983.

Willer, H. 1985. Zukunftschancen nachwachsender Rohstoffe. *Zuckerind.*, 110-62.

3

BIOTECHNOLOGY AND HUMAN ENVIRONMENT

A. BLAZEJ
Slovak Technical University, Bratislava, Slovakia

Abstract. As a result of the growing awareness of environmental problems, the UNESCO-ROSTE European Workshop and Training Course in Modern Biotechnology 1991 held in Prague was devoted to the relationship between biotechnology and the human environment. We are entering a new age, the age of science, hi-tech, and science-based industry. We are also entering the age of 'Environmentalism'. Prosperity in the near future particularly in the developed industrial countries depends on how well we manage to create a new kind of socio-economic development, a new living style, a new 'consumer culture' based on rational consumption of goods, new non-waste technologies of industrial production, and high environmental quality of products.

Biotechnology is in a position to take up these challenges and positively influence and improve the quality of human life. In this paper, the following themes are addressed:

- conception of human environment
- conception of biotechnology
- environmental biotechnology
- biologization of national economy.

Biotechnology can contribute to solving global problems of mankind and global environmental problems, and can revolutionize many sectors of industry. The impact of biotechnology on the chemical industry, agriculture, and food production is presented.

INTRODUCTION

In the past 50 years, worldwide tendencies have been directed towards:

- speeding-up economic and social development
- stimulating material consumption
- ensuring people's material wealth.

Emphasis has been and still is placed on economic growth. The cult of economic growth is deep in the social consciousness as the main source of well-being of an individual as well as of society.

Economic thinking and the long-term economic strategy for future years in many countries are based on the theory of material production growth. This also results in an increase of innovative activity and both scientific-technological and trade policies are subordinated to this fact.

Existing economic development is accompanied by an immense squandering of raw material and traditional energy sources, and has brought concomitant high ecological stress due to production and consumer waste. Man's value orientations were in essence anti-ecological, connected with overproduction and overconsumption in the highly industrialized countries. The distribution of wealth proceeds unevenly. The Third World remains ill-fed, ill-clad, ill-housed, and illiterate. Our present task is to face and solve this dilemma.

Industry is central to the economies of modern societies and is the indispensable engine of growth. Many essential human needs can be met only through goods and services provided by industry. There has been remarkable technological progress since the industrial revolution.

In many fields, extraordinary technological breakthroughs have been achieved and others will follow. In fact, one of the great paradoxes of technology is that principally it has contributed to human progress, modern civilization, and human wealth, but has caused the degradation of the biosphere connected with the problem of the real existence of man and his survival.

Currently, mankind is facing some global problems:

- environment
- food security
- raw materials
- population growth
- energy resources
- sound economic and social development.

In fact, a greater threat to international security lies in the continuing degradation of the biosphere. The main global environmental problems are:

- pollution of air, water, and soil
- global warming
- depletion of ozone layer
- acid rain
- deforestation
- extinction of biological species.

Biotechnology can contribute to solving all global and global-environment problems.

HUMAN ENVIRONMENT

The human environment is a complex multicomponent system created by (Blazej *et al.*, 1981):

- *natural environment*: physical, chemical, and biological composition of nature, where man is in interaction with his natural surroundings.
- *man-made environment*: artificial material environment constructed and created by human activities.
- *working environment:* man-made or combined natural system and man-made material components in industry, agroforestry, and services with human economic activities.
- *socio-economic environment*: where man realizes his biological, material, social, societal, and cultural needs.

All these interactions are shown in Fig. 3.1.

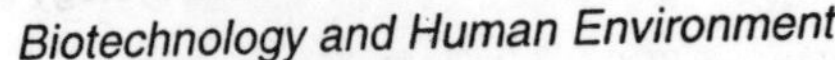

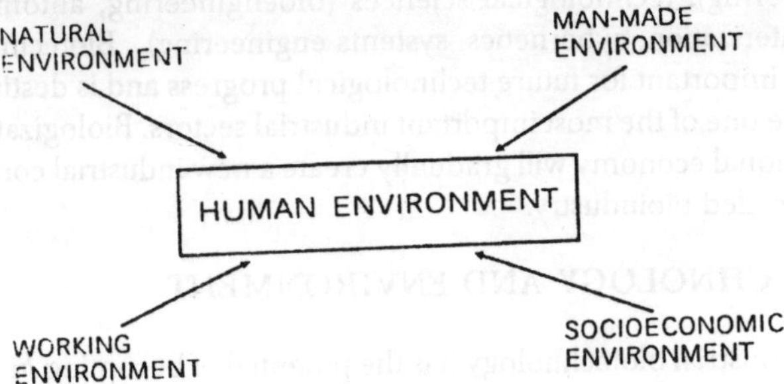

Fig. 3.1: Human environment.

The real progress of society can be identified only with an increase of quality of life. Development of society must be directed at a new quality of socio-economic development, where human environmental quality is a primary factor to which economic growth is subordinated. This means that a healthy human environment is the basis, and economic activities must be derived from it. The human environment becomes the main component of quality of life and therefore must have priority over all other human priorities. The quality of human life is created by these components:

- quality of human environment
- life style
- material wealth
- social security
- human relations.

CONCEPTION OF BIOTECHNOLOGY

There are many different definitions of biotechnology. It was said by Rose that there are as many definitions of biotechnology as the number of people anxious to be known as biotechnologists. The tendency is that all practical exploitations of biological knowledge for the benefit of man are termed biotechnology. Instead *industrial biotechnology* can be defined as technology that uses biological organisms, their components, and biological processes in the *man-made* environment for industrial manufacture. Industrial biotechnology is integrated technology that uses scientific and technological knowledge from biological, chemical, and technological sciences.

Advanced or modern biotechnology uses knowledge from molecular biology (genetic manipulation), enzymology (protein engineering), technological sciences (bioengineering, automation, computerization, cybernetics, systems engineering) . Biotechnology is very important for future technological progress and is destined to become one of the most important industrial sectors. Biologization of the national economy will gradually create a new industrial complex, the so-called bioindustry.

BIOTECHNOLOGY AND ENVIRONMENT

Industrial biotechnology has the potential of becoming *hi-biotech* because it will solve environmental problems like the organic part of production. Hi-biotech is based on:

- science- based industry
- high-technology
- non-waste production.

The positive effects of biotechnology on the environment may be summarized as:

- it uses renewable raw materials
- it can be based on complex utilization of raw materials
- it provides innovation or alternative processes to conventional technologies which are environmentally sound
- it produces biodegradable products and by products
- its products and goods have higher environmental quality and better ecological properties
- a higher index of recycling of consumer goods can be attained
- it supports sustainable socio-economic development.

At present, we have to accept that biotechnology is not *a priori* green technology or hi-teach, because it has negative effects on the human environment despite its advanced nature, like modern technology.

The main disadvantages of current modern biotechnology are:

- incomplete utilization of C-substrates and nutrition solutions
- low concentration of cells, organelles or biocatalysts in biochemical processes
- low concentration of products
- formation of by-products or wastes
- high cost of biological and technological limitations
- downstream processes yielding excessive waste and consuming excessive energy with regard to realizing optimal yields of final products

- the spreading of pathogenic micro-organisms into the environment with risk to man, animals, and plants
- the release of rDNA or micro-organisms with rDNA into the environment (undesirable horizontal gene transfer to other organisms).

ENVIRONMENTAL BIOTECHNOLOGY

Environmental biotechnology represents a new area of environmental sciences based on biochemical, biological, and biotechnological approaches to solving of environmental problems. The major areas are:

- the biotechnological elimination of environmental pollutants (gas, liquid, solid) produced in any sector of the national economy
- depollution biotechnology processes like alternative processes to conventional non-environmental ones
- biosafety in biotechnology
- biological quality control of the human environment.

It should be stressed that a qualitative new situation is created when micro-organisms are employed with recombinant DNA, especially with pathogenic properties. Very extensive and highly controversial discussion has revolved around the possible dangers in this case. Most countries have legislation relating to health and safety at work which promotes, encourages, and sets high standards of working conditions. The objective is to protect both the people at their workplace and others against risks arising from their actions. In some cases, there are regulations to cover the keeping and use of dangerous substances, as well as the emission of noxious or offensive substances into the environment.

The WHO and the EFB party of Safety in Biotechnology are introducing a 4-class system in biotechnology which assigns the individual micro-organisms used to certain risk categories (Hubber, 1991).

The EFB classes are:

I harmless
II low risk
III medium risk
IV high risk.

Many of the harmless micro-organisms of class I have a long history of safe use in large-scale biotechnological production (antibiotics, enzymes) as well as in human consumption (beer, wine) or in agriculture (biopesticides).

Class II organisms (low risk) are weak pathogens, usually employed in the production of vaccines or in diagnostics. The extent of their use (compared with class 1) is fairly limited.

Hardly any practical use is made in industrial production of medium risk (class III) and high risk (class IV) organisms.

There is generally worldwide agreement that this 4-class system in industrial biotechnology is sufficient for practical purposes.

As far as rDNA organisms are concerned, there is now a more or less general consensus of opinion that:

- no fundamentally of qualitatively new risks are to be expected
- rDNA and rDNA organisms cannot be classified *per se* as pathogenic or hazardous
- the transfer of plasmids and genes is a natural process which occurs in every habitat
- reliable risk assessment is possible if the properties of the donor, recipient, and vector are known.

The WHO accepted in 1983 the view as follows: "*There are no unique or specific safety risks associated with recombinant DNA work (genetic engineering). The risks are no greater than those associated with work where use is made of known pathogens, and they do not necessitate special laboratory design or practice*".

BIOLOGIZATION OF NATIONAL ECONOMY

In this paper, the term 'biologization' of the national economy is used to indicate the trend of utilization of biotechnology as the new economy of any country. We can look at the expansion of biotechnology in many sectors of industry. The many expectations for the 21st century can potentially be realized through biotechnology. I personally believe that biotechnology can also contribute to solving the main global problems (Table 3.1) and environmental problems (Table 3.2) over the coming decades. For illustration, the impact of biotechnology on the chemical industry, agriculture and food production is presented.

TABLE: 3.1
CONTRIBUTION OF BIOTECHNOLOGY TO SOLVING THE MAIN GLOBAL PROBLEMS

Global problems	*Position of biotechnology*
Peace and security	Biological weapons (these must be destroyed)Environmental security can be solved through biotechnology
Environment	Environmental biotechnology
Food security	Agricultural biotechnology Microbiological production of food
Energy resources	Biofuels–biogas, methanol, ethanol, hydrogen (by bio-photolysis of water)
Raw materials	Biomass (phytomass, zoomass, microbial mass) Renewable raw materials
Population growth	Birth control, bio-health care Bio-pharmaceuticals
Appropriate economic and social development	Contribution to sustainable socio-economic development through biologization of national economy

TABLE: 3.2
CONTRIBUTION OF BIOTECHNOLOGY TO SOLVING THE MAIN ENVIRONMENTAL PROBLEMS

Global environmental problems	*Position of biotechnology*
Pollution wastes	Depollution technologies based on biological, biochemical, and biotechnological processes
Global warming	Bioelectrocatalytic reduction of CO_2 Acceleration of carbon cycle in nature by man made photosynthesis
Acid rain	Biotechnological desulphuration of fuels Biotechnological denitrification (removal of NO_x)
Deforestation	Genetically improved higher plants
Extinction of biological species	Through biologization to improve the quality of life on the Earth

BIOTECHNOLOGY AND CHEMICAL INDUSTRY

The chemical industry has made great progress over the last 50 years and has contributed considerably to the well-being of mankind (health care, food, clothing, footwear, chemical materials, fuel, etc.).

In the last decade, the chemical industry has invested heavily in facilities for environmental protection, with visible success. Nevertheless, the chemical industry has some obstacles, for instance:

- raw-material-intensive, organic technology is based on fossil raw material (coal, crude oil, and natural gas)
- energy -intensive (uses for technological processes, high-temperature and high pressures)
- capital-investment-intensive(uses sophisticated technique or production, high safety responsibility during manufacturing, storage, transport, use, and final disposal)
- environmentally not fully acceptable (chemical production risk for the environment is high and products are usually toxic and xenobiotics).

The human environment is overloaded by chemicals and chemical materials. The number of individual chemicals is about 4.5 million. A present time, commercial chemical products represent about 100,000 chemicals. Each year about 1000 new products are added to the market. The main sources of chemical pollution are chemical processes based on incomplete chemical reactions that depend on the state of chemical equilibrium. A final product, reaction substances, side products and impurities need separation, isolation, and purification. All these operations produced pollutants. Chemical phobia among the public and a fear of chemical carcinogenity, plus non-biodegradable chemical substances and chemical ploymeric materials, have caused a strong rejection of the chemical industry. Pollution problems often complicate with strong odour and colour of chemicals even in very low concentration.

The second main source of chemical pollution arises from the application of chemicals and chemical materials as well as chemical processes in other sectors of the national economy. Pollution problems of this kind are very often caused by a misapplication of chemicals and chemical materials by users.

Biotechnology can contribute to the solution of many chemical problems through the biologization of chemical technology. A sound future lies ahead for hybrid technologies (combination of chemical and biochemical processes). Chemical and biochemical reactors are key equipment in both industries. Similarities and differences between chemical and biotechnological reactors are summarized in Table 3.3 (Bales, 1990). Comparison of products manufactured by the chemical industry and industrial biotechnology are shown in Table 3.4.

TABLE: 3.4
SIMILARITIES AND DIFFERENCES BETWEEN CHEMICAL AND BIOCHEMICAL REACTORS. BALES, 1990

Characteristics	*Chemistry*	*Biochemistry*
Reaction conditions	Often extreme (p,T, pH)	Very modest (P, T, pH)
Concentration	High	In most diluted water solutions
Rate constant	s, ms^{-1}	min, h^{-1}
Continuation	As a rule	Only occasionally
Type of reactors	Type construction	Mostly individual construction
Catalysis	Chemicals catalysis, extreme conditions	Biocatalysis, very modest conditions
Automation control	Simple measurement controlled by computer simply, rapid response	Direct control by computer only in early stages, slow response
Risk for environment	High toxicity and xenobiotics	Biological processes do not threaten environment
Energy requirements	Energy intensive	Less energy-intensive
Water requirements	Large problems with purification of waste water	Also large problems, but better biodegradability and a recirculation of water at lower cost

TABLE: 3.4
COMPARISON OF PRODUCTS MANUFACTURED BY CHEMICAL INDUSTRY AND INDUSTRIAL BIOTECHNOLOGY

Product	*Chemical industry*	*Industrial biotechnology*
Raw material	Fossil	Organic C-source (renewable)
Organic chemicals	Synthetic organic chemicals	Natural products (primary secondary metabolites) Microbial biochemicals
Chemical materials	Synthetic polymers	Biopolymers (natural and microbial)
	Plastics	Bioplastics (biodegradable)
	Fibres	Biofibres (better hygienic properties)
	Rubbers	Biorubbers
	Composites	Biocomposites

BIOTECHNOLOGY AND AGRICULTURE

Plant products derived from fewer than 30 plant species provide more than 90% of human food. Eight cereal crops supply more than half the world's demand for cereals. Animal products contribute over

60 million tonnes of edible food proteins. We can observe the increasing importance of marine food products. The potential of microbiological production of food (e.g., SCP for human feeding) is not exploited.

The chemistry of 'chemization' of agriculture has contributed to the intensification of agriculture production. Its ratio on increasing of the crop yield is about 50% (industrial fertilizers, pesticides, plant growth hormone). Chemization of agriculture has resulted in serious environmental problems, such as:

- high soil salinity
- heavy metals and metalloids in soil
- low soil acid pH
- deficiency of organic carbon in soil
- eutrophication of surface water
- pollution of underground and surface water
- chemical stress of plants
- lower environmental quality of food.

Contrasting with this negative impact of chemization on agriculture production is the biologization of cereal production. Potential and average yield of two major crops in European developed countries are shown in Table 3.5.

TABLE: 3.5
AVERAGE AND POTENTIAL YIELD OF TWO MAJOR CROPS IN DEVELOPED COUNTRIES OF EUROPE

		Yield, tonne per hectare	
		Wheat	*Maize (corn)*
Potential Yield		*14*	*22*
Average yield			
	Years		
	1960-65	1.7-2.0	3.5
	1980	2.5-2.8	5.8
	1990	5.9-6.1	7.1

Biotechnology can effect deep changes in agriculture in the area of plant and animal production:

Plants

- gene manipulation of crop plant cell in order to increase productivity and the value of the crop yield and crop quality

- maintaining or increasing yield by increasing resistance of plants to pest, disease, drought, response to fertilizers, etc.
- improved storage quality and processing properties
- in the agriculture seed industry
- production of biofertilizers (bacterial fertilizers, biocomposts, biopesticides, plant growth hormones, N-fixation by plants)
- genetic manipulation of soil bacteria that promote or impede plant growth can also be carried out
- increase of plant health care and decrease of plant diseases.

Animals

The contribution of animal biotechnology includes:

- the genetic improvement of animal breeds
- improvements in the reproductive efficiency of livestock
- increase in animal health care (antibiotics, vaccines, monoclonal antibodies, diagnosis, prevention)
- up-grading of animal feeding and new fooder (proteinization of straw, SCP, amino acids, vitamins, antibiotics, enzymes, growth hormones).

BIOTECHNOLOGY AND FOOD INDUSTRY

The first industrial utilization of biotechnology has been practised in the food industry for several thousand years (brewing, baking, fermented foods). No-one had any knowledge of the existence of micro-organisms or biological processes till the Pasteur period. Despite advances in biological sciences, the food industry remained more or less an empirically-based industry using 'low-tech' processes based on experience and art rather than being a science-based industry.

The impact of modern biotechnology in the food industry is lower than in agriculture. At present it is necessary to formulate achievable objectives to aid the rational improvement of food production and processing technologies and to reach desired goals for food product quality and safety.

Biotechnology oriented towards the food industry can bring:

- significant increase in food quality
- a higher nutritional value of food
- improved traditional and complex new processes
- provision of ingredients
- lower energy consumption

- higher added-value products
- less waste
- economic benefits
- complex utilization of food raw materials
- improvement of food-grade micro-organisms (bacteria, yeasts, fungi)
- use of more rDNA technologies and fundamental understanding of micro-organisms useful in food fermentation and preservation processes.

A new area is opened for microbial food product, for example:

- SCP for human diet
- food enzymes
- essential amino acid production
- organic acids
- SCO (single cell oil)
- essential fatty acids
- new type of additives
- microbial polysaccharides
- new kinds of sweeteners.

CONCLUSION

People are now beginning to realize the urgent need for a fundamental revision in the relationship between man and nature. A global environmental crisis now faces mankind and the planet Earth. The time has come to resolve these crises in the 1990s with a view to the survival of all life on Earth. We can succeed in ensuring environmental security. A series of actions in international, regional, and national frameworks could be taken (Blazej and Privarova, 1991).

First, it must be recognized that in the developed world, the overconsumption of goods and services must decline and a new consumer culture is needed. On the other hand, poverty in the developing world is one of the greatest threats to the global environment. The central problem is in reducing overproduction on the one hand, and removing poverty on the other.

Second, we must respect the differing environmental priorities of industrial countries. One may have a shortage of drinking water, while the other is more concerned with water pollution, and still other countries are concerned with rapid soil degradation or air pollution, etc. A unifying concern for the environment does not dictate a search for uniform environmental priorities or uniform environmental standards. But the global environmental crisis must seek to integrate differing national perceptions, sensitivities, and priorities within a new framework of co-operation in environmentally sound development.

Third, the vision and perception of our common future exists in a wide-ranging agreement that sustainable development strategy should meet the needs of the present generation without compromising the ability of the future generation to meet their own needs (Brundtland, 1987). The concept of sustainable development has not yet been rigorously and acceptably defined. It must include the protection of future economic growth rates as well as future human development levels.

Fourth, these realities in the human environment require people to re-think. They must adjust their behaviour to the environment and formulate a socio-economical system that focuses as much attention on people as it does on technology.

We conclude that immediately targeted reductions of all types of pollutants are required. Biotechnology can make an essential contribution to solving many global, regional, and national environmental problems over the coming new century.

REFERENCES

Bales, V. 1990. Chemickoinzinierska prax e vyskum v novych ekonomickych podmienkach. *Zbornik 17 Celostatnej Konf. OSCH190.* 1 Cast Tale.

Blazej, A. and Privarova V. (eds). 1991. *Environmental Biotechnology.* Elsevier.

Brundtland, G.H. 1987. *Our Common Future.* Oxford University Press.

Hubber, L. 1991. Positive and negative impacts of biotechnology on the environment. in A. Blazej and V. Privarova (eds), *Environmental Biotechnology,* Elsevier.

4

BIOTECHNOLOGY AND ECONOMICS

B.K. ZIMMERMAN
Spectrum International, Finland, and Renvall Institute, Helsinki University, Finland

It is tempting at conferences such as these for us to become immersed in the challenges of the technology and the underlying science. To be sure, these are exciting times; the nearly limitless potential of technology based on life processes is only just beginning to be realized and we are all very eager to help its advance into the future.

But the other side of biotechnology–the 'business side' if you will–is of no less practical importance than the scientific and the technical. It is the interface between the isolation of the laboratory and that society which it purports to benefit. And it is ultimately the major source of fuel needed to drive the engine of research and development. In other words, it is where the money comes from.

This paper, then, may be considered a discourse in applied economics, with specific reference to biotechnology. I would also like to consider the transitional economic picture of East Europe, and the standing of a biotechnology industry in redevelopment.

BASIC ECONOMICS

To put it very simply, economics is the science of value, both real and perceived (although some economists would argue that the perception of value is all that really matters). It is involved with the creation of value, including the value of labour, the production and exchange of goods that contain value, and the currency that we exchange and save to represent value.

Thus one who happens to stumble upon a gold nugget weighing in at five kilos has discovered a great deal of value. But it is only so valuable because overall, gold is a very rare resource, that usually requires a lot of effort (investment) to obtain in pure form. Finally, it is of high value only because people value gold, for a variety of historical and chemical reasons. This man, like the one who wins the state lottery and received ten million of whatever currency units, is the lucky exception.

For most, the economics equations are rather well balanced, in terms of putting in the value of labour, investment money, time, scientific creativity, artistic talent, etc., and getting out value of a different kind, hopefully exceeding the value that was invested. The statement, when developed into a semi quantitative science, subsumes a major portion of the field of economics.

Thus, the man who buys a piece of barren land very cheaply may be able to reap considerable created value in forty years, if his land is covered with a forest of tall, straight timber after that time. He has invested very little money and very little labour, but a great deal of time.

On the other hand, the farmer who invests his labour in planting potatoes every year will work much harder than his forest-owning neighbour, will receive less from each crop than if it were timber, but he will receive it every year, and he will have in hand the value, created from his labour, that he needs to conduct his life.

Another of great talent may have a flash of brilliance resulting in an invention which he can then patent and sell for a high price, representing an equivalent perceived value of his invention. In this case, his invention represents 'new' value, in that when it enters into worldwide commerce, the total of the money paid for the use of his invention, minus the costs incurred, will be substantially greater then zero. While not all brilliant inventors are appropriately rewarded (perhaps they forget to file a patent, or foolishly sell their idea for far less than its true worth), some collection of people will profit handsomely.

THE ENTREPRENEUR

As is often the case, however, it is the worldly individual with a nose for business who will fare better than the often unworldly inventive genius. It is the entrepreneur–the person with a sense of what will bring him the greatest increase in new or created value in the shortest time–who may be the one to reap the big gains. Thus it is often such an entrepreneur who will spot the creative inventor–or scientist–and have a sense of the ultimate economic value of the inventor's creative efforts. Thus by enabling the invention to become available to a large number of users who will pay for its use, the entrepreneur has not only performed a valuable service to society, he has also done well for himself in the process. Why, then, aren't there more entrepreneurs?

True entrepreneurism, like musical talent, is not a gift that is universally distributed. But perhaps more important, the entrepreneur also takes many risks, often risking his personal wealth (value) in the hope of increasing it. Many times the entrepreneurs do not succeed, often for reasons well beyond their control, and because of the unpredictable fluctuations and perversities of human life and behaviour. Risk taking, especially on a grand scale, is not for everyone. Most prefer the comfort and security of a steady job, where they know that their share of the value created by their own labour (minus, of course, taxes, overheads, and company profit), is sufficient to provide for their own needs and those of their family.

Nevertheless, for a dynamic economic society, at least a small cadre of entrepreneurs, who must be both bold and clever, is essential. For these individuals are the source of innovation in business and commerce, and the driving force behind the development of new and more advanced products in all areas. Just as the brightest scientists are the source of the fountain of new knowledge that continues to advance our understanding of the universe in which we live, the creative entrepreneurs are absolutely essential to prevent economic stagnation. This is why most of the advanced, new technology comes not from the large, well established (and, hence, conservative) corporations, but from the new and adventurous companies, willing to take chances and create revolutions. On the other hand, as we have witnessed during the past 70 years, social and political environments that stifle the entrepreneur lead to economic suffocation.

TRADITIONAL VERSUS HIGH-TECH BUSINESS

The business based on the development of innovation is generally subject to a different kind of economic model from that governing 'traditional' business. That is, the manufacture making shoes, and selling potatoes, must sell shoes or potatoes for at least as much as it costs him to produce them. The 'profit margin' is the difference between revenues taken in, and costs expended, usually expressed as a percentage of total revenues.

Figure 4.1 illustrates a typical business in which a small but steady profit margin is maintained, in the face of steadily rising costs. Figure 4.1 (a) represents the case of a more or less constant profit in absolute terms, but a declining profit margin. Figure 4.1 (b) would represent a more desirable situation, in which the profit margin remains fixed, but the amount of profit increases to reflect general increasing costs. The increasing annual costs themselves could in turn reflect a number of factors, including inflation, higher taxes, or a business expanding in volume. The profit margin, to a first approximation, will represent the added-value of labour. One can, of course, extend the arguments to include mechanized labour, taking into account the investment in labour-saving machinery and technology.

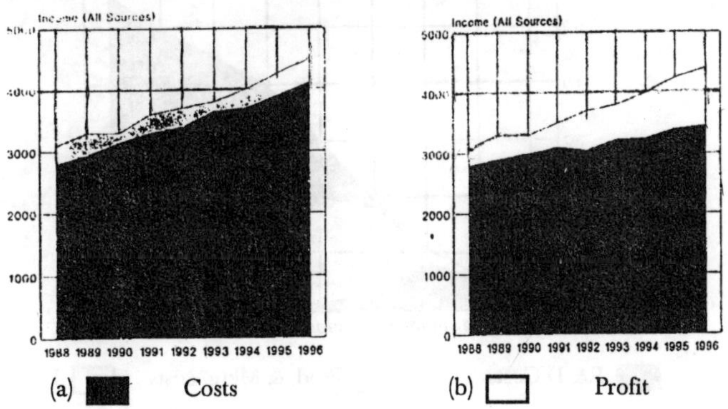

Figure 4.1. Traditional model -selling products, services.

The entrepreneurial, high-tech model of business differs in several significant features. First, considerable time and costs will be required before there can be any income from sales. Thus the company developing the product must have significant sources of financing to meet these costs.

The second major way in which this kind of business differs from the traditional manufacturing enterprise is that the 'added' value of the product is usually very high. That is, the price for which the product is sold will represent far more than the sum of the raw materials, plus the added-value of the labour and/or manufacturing costs required to produce the product. It will also reflect the investment in R & D, as well as the premium demanded by true innovation. The price will also be highly dependent on the product's uniqueness and whether or not there are competitors selling a similar product. For that reason, patent protection for intellectual property takes on a much greater importance than it does in producing generic items, where the price is determined primarily by supply and demand.

Thus, the so-called 'profit margin' for such innovative products, once sales begin (which may be many years after development has begun), defined as a percentage of sales, can be very high. Figure 4.2 represents an ideal high-tech product, where after five years or so, annual profits are high. At the same time, during the early years of development, the company operates at a loss, and must, therefore, secure financing to cover the often very high development costs.

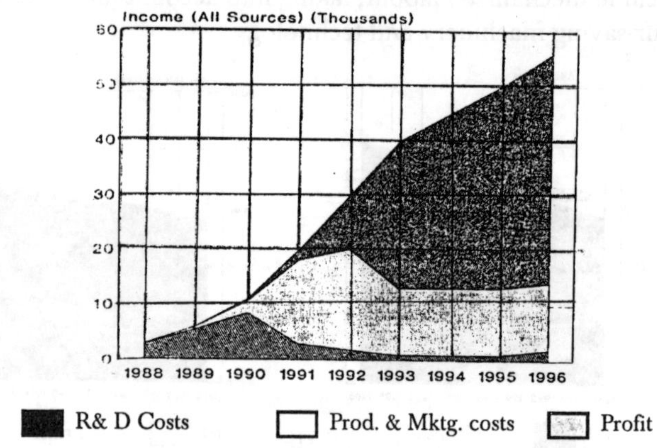

Fig. 4.2: High-tech value-added model-creation of high-value technology and products.

Where does an entrepreneur with a good product idea find the capital necessary to finance a company for what can be a very long 'dry' period? This is probably the single most important aspect of applied business economics, which will be treated in detail later in this discussion.

THE ECONOMICS OF BIOTECHNOLOGY

While it would be possible to continue on in a general fashion, it is much more to the point to continue this discussion from the point of view of the nascent 'biotechnology industry', if you permit me to use the term (some would not).

Let us first consider the requirements for a successful biotechnological enterprise. What does the entrepreneur in biotechnology have to know? On what will be build his business? How will the money to get started be obtained? How long will it take before his enterprise has created 'value'? when will he have something to sell? When he does, will anyone buy it? And how about his competitors, who are keeping their activities a big secret?

The following is a summary of the main features for the would-be entrepreneur of a new biotechnology company to consider very carefully.

THE 'BUSINESS IDEA'

First of all, what is the concept around which one wishes to create a business enterprise? More specifically, what is (are) the products(s)? In general, one is more successful the simpler the initial concept. That is, and especially in biotechnology, focus on a single technical area or 'niche', in which one may aspire to be the best, and develop unique products for which there will be little or no competition, is the key to a winning strategy.

Jumping on bandwagons, where the initiative has already been seized by somcone else, usually does not work. For example, in the early days of the pioneering biotechnology companies (in the late 1970s and early 1980s), the magic word was 'interferon', seen by some to be the long sought cure for cancer, and then some. Of course it was not. The scientists who really knew what was going on were well aware, even in 1975, that the interferons, while they might be of some therapeutic value for cancer and viral diseases, probably in conjunction with other therapies, were never going to be magic bullets, and certainly did not represent the final solution to the cancer problem. But they were not altogether open about it. It was a target protein that could be made by genetic manipulation techniques in heterologous hosts. And the fledgling industry needed money–badly.

Thus the hype that surrounded the interferons, and some other prospective products, worked for a time on the professional investors looking for a quick route to as much added-value for their dollars as possible. Even the public could be heard mumbling about these new wonder compounds that would cure cancer. Drawn in by the rampant hype, the first public offerings of stock in biotechnology companies (Genenetech, in 1981, followed a few months later by Cetus) raised many millions, and were fully subscribed almost at the time they were offered, leaving thousands of disappointed would-be investors empty handed. There is of course a risk in over-subscribing a public offer. When the dust has settled, the value of the shares, multiplied by the number of outstanding shares, should be a reasonably realistic representation of the value of the company. If the company value is too low, because the offer was undersubscribed, or sold for too low a share price, the trading price will rise. That, however, will not bring much needed cash into the company's treasury.

On the other hand, if the number is obviously much too high, as was the case following Cetus' record offer, the price of the shares will fall. In Cetus' case, the share price remained at one half to one-third of the offered value for several years following the public offering, which had a definite chilling effect on both public and investor confidence in the company.

Apparently, at least one national government (Cuba) was also taken in by the furor emanating from the environs of San Franscisco, and launched a massive national biotechnology programme, and built a large research centre, with interferon to be the prime focus.

There was, of course, a price to pay. Interferon has long since lost its lustre, and the investment community took the scientific entrepreneurs to task for overselling their wonder drug, or at least failing to make good on their promises.

On the other hand, if interferon had turned out to be as good as some claimed, trying to play 'catch-up' would have been a futile endeavour. The business idea must be original, it must offer something new and unique to technology, it must be something that people need, or at least want, and are willing to pay for, and you should be the first one to have and run with the idea. Of course, as the applications of biotechnology spread in every direction, one hears complaints that it is increasingly difficult to come up with a novel business idea and /or products. I disagree. The rate of discovery and of innovation is, in my opinion, still in early log phase, compared with the size of the body of knowledge that would have to be learned before the scientific enterprise begins to slow down. True, the number of applications is increasing exponentially, but so are the scientific bases on which such technology is founded.

INTELLECTUAL PROPERTY

O.K. So you have convinced yourself that your business concept is fantastic. Do you have access to the technology needed to develop the idea? If so, is your route of access protected, so that others cannot jump on your coat-tails, or worse yet, take your ideas and develop them first?

If you are the inventor, patent your invention. If you are working with someone else who has made a brilliant discovery that has far reaching commercial potential, be sure that the information is properly protected.

Usually this will mean that you enter into a legal agreement with the inventor, paying the costs of worldwide patent preparations and filings, plus appropriate compensation for an exclusive licence to develop the invention commercially, plus royalties on sales that may be forthcoming.

RESOURCES

Now that the entrepreneur can guarantee access to a unique, winning technology, what does he have in hand to develop it with? First, he must have or have access to scientific and technical resources, meaning some first class scientists and technologies who are willing to work on developing a product, which for them may be less glamorous and intellectually rewarding than a career in basic research.

This sounds as though it should not be such a difficult requirement to meet. But in fact, it has not been at all easy, even for companies that have had ample financial resources to pay high salaries and provide state-of-the art laboratories. The best and most creative scientists usually prefer the freedom of a university laboratory or a research institute where they can pursue any avenue of investigation that looks interesting. In a corporate environment, where the achievement of a specific goal must be obtained as quickly and efficiently as possible, such freedom must necessarily be constrained, although companies must be careful not to try to channel the efforts of scientists too narrowly, since originality and creativity at all levels of development are important. And some very important innovations have been made by corporate scientists (for example, the ploymerase chain reaction (PCR) was developed by Cetus scientist Cary Mullis). Nevertheless, the corporate scientist is not nearly as free to publish and talk about his work as the one who does not have to worry about competitors, patents, and profitability. In any case, in terms of the economics of job markets, all of this means that you will probably have to pay a scientist somewhat more or work for your company than he or she earned at a university or research institute. It is worth noting that some first-class scientists were offered top managerial or supervisory posts in some of the early biotechnology companies, often with disastrous results. Talent in the laboratory does not imply organizational or managerial talent, or the critically important skills of dealing with people.

So much for the human resources. There are other kinds of course. Reagents. Raw materials. Proper laboratory facilities. These must be available at an affordable price.

FINANCING

One condition that the entrepreneur must meet in no uncertain terms necessary, if not sufficient, for a successful business, is to have a large enough financial base to begin with. Without it, one risks running out of money, or at least not being able to make longterm commitments to one's staff. But people, even scientists, like security, not uncertainty over their futures.

There are of course, several possible sources of early financial support. The importance of venture capital has been discussed above. But for some very original 'far-out' ideas, that may be rather hard to come by. Of course, one can use personal funds or, if one is fortunate, persuade a rich relative or two to invest in one's ideas (don't laugh–a number of successful biotechnology companies have been started in precisely this way). If one can establish good credit or convince one's banker that one's business idea will pay off in spades, it is sometimes possible to borrow at least a part of the required capital. The problem is, you have to pay it back. Since start-up biotechnology companies are, almost by definition, risky propositions, there is always a chance you will not be able to pay it back and, hence, will lose your shirt.

It is really better if you can take the inherent risks in the biotechnology business with somebody else's money. The professional venture capital companies, if they really believe in you and your ideas, and your ability to realize them, can provide you with all the money you need.

There are in the United States, Europe, and the Far East, a concentration of venture capital companies, those that specialize in high risk/high-profit new ventures. If the venture capitalist is convinced that the technology is sound, the business idea exciting, and the risks reasonable, he may stake a new company for several million dollars in start-up financing. For that, he will want a significant fraction of the corporate shares, or ownership of the new company, and may impose as a condition for investment, a seat on the board of directors, or the right to choose the corporate managers.

But the venture capitalist does not usually plan to wait until the product is on the market and making a big profit, or indeed any profit at all. The first question usually asked is "*When can I get my money out?*". This really means "*When can I get at least ten times my original investment back?*" which in reality may mean "*When are you going public*".

ADDED-VALUE

Thus we come to another meaning of added-value. This is the added-value of a company developing exciting new technology *as perceived by the investors.* This is the basic principle of entrepreneurial economics that enables a company with unproven ideas to succeed, even when it may burn up a hundred million dollars before it has even sold a single product. In fact, of the several hundred or thousands of so-called biotechnology companies that have been started since The Beginning, only a handful have been profitable or are now making a profit.

How can that be? Is not the fundamental law of good business that you must take in more than you spend? Apparently not in biotechnology. Here is the heart of why an advanced biotechnology company operates on totally different principles of economics than, say, a dairy, or a manufacturer of bricks, or a potato farmer, or a brewery, or a construction company.

All must create value. But in the traditional business, the value of one's labour is added to the value of the raw material, to create a product of higher value, for which people will pay a price that exceeds the outlay by the company. One way of viewing the profit then is to consider it the added-value of labour.

In a company where the focus is on developing advanced new technology, its apparent or trading value is based not on current reality, but on expectations of future value. That is, the tangible assets of a small biotechnology company will generally be very small in comparison with the amount of money needed to develop a product. And each year the product is under development, the costs will continue to rise. And for example, if a new pharmaceutical or vaccine ever reaches the stage of human clinical trials, the costs will rise even more, still without a guarantee that there will be a sale-able product at the end of the road.

Yet the company engaged in the development of such a product has considerable value. That value is determined simply by what such a company could be sold for if it were to be offered for sale. In general, such companies bring a price that substantially exceeds the total cash and other tangible assets, even including the market value of patents held. The difference is the added-value returned on the total investment. This added-value can be very large, even when the company is and always has been operating at a significant loss every year and has never sold a single product. Why?

Simply because of the perception that at some point in the future, there will be technology or products based on that technology that can be sold on the open market for real money that will be in substantial excess of the total investment to date.

HOW RISK FINANCING WORKS

It is perhaps best to illustrate how 'deficit' financing works, ultimately to everyone's benefit, with a hypothetical example.

Let us suppose that a scientist and a businessman team up to found a company to develop a revolutionary genetically-engineered AIDS vaccine. They put up us$ 100,000 of their own money to register a company (the legal minimum capitalization may vary from trivial sums to up to about us$ 50,000, depending on the country, but a substantial 'good faith' investment by the founders is usually needed to establish their credibility with investors).

In the beginning, when the idea is little more than a laboratory finding and the investment needs great, the risk will be seen as high. Thus, the initial, or start-up shares, will have a relatively low unit price. But the scientist's scientific strategy to create an AIDS vaccine is revolutionary and backed by some preliminary experiments in his university laboratory. The founders therefore succeed in attracting an initial investment of two million dollars, for which they will sell 20% of the shares. By definition, their business concept now has a 'market value' of US$ 10 million, and the paper value of the founders' shares is now US$ 8 million, substantially more than the US$ 100,000 they spent to start the company.

After one year, they have used up one million and are well into the second, but they have also established the validity of their idea, and tested a prototype on animals. Thus their business concept has become less risky, and its perceived value is therefore higher. Thus when they now wish to raise additional money to cover further have costs, they are now able to sell an additional 20% of the company's shares for US$ 5 million. Since that transaction, in effect, defines the value of the company and the value per share, the company now claims to be worth US$ 25 million, which is the current crude 'expectation value' of the product being developed, compared to an evaluation of US$ 10 million at the time of the first financing round.

The original investor is now quite happy, because the shares for which he originally paid US$ 2 million now have a market value of US$ 5 million. Thus, in one year, the 'added-value' of his investment is US$ 3 million, reflecting progress toward validating that the original exciting idea was sound after all.

But such 'private placements' generally are strictly controlled; the investors are not free to sell their shares on the open (stock) market. In order for such early-stage investment in such a fledgling company to be interesting, there must come a time when there will be either a public offering, at a much higher price per share than those paid by the early-stage investors, and following which shares can be freely bought and sold, or a private placement to a major institutional investor, to which the initial shareholders will be allowed to sell shares. 'Going public' is the preferred exit route for the venture capitalist. At such a time, shares may sell from ten to (in extreme cases) 1000 times their original value, the short-term profits to the venture capitalist being enormous.

This mechanism, by the way, is the origin of the feature 'Biotech Millionaires' appearing annually in *Genetic Engineering News.* Most of these are founders or early-stage investors in biotechnology companies that have continued to excite the imagination of the financial community, as well as (by now) the scientific and medical community. The founders, many of them research scientists, probably had to borrow their meagre start-up capital against their home equity or from relatives originally, but in a relatively short time held stock with a paper value worth millions. It should be kept in mind, however, that starting a biotechnology company is scarcely a quick and dirty way to untold riches. For that to happen, (a) the promises so eloquently and persuasively set forth in the original business plan must continue to be realistic, or be replaced by even more promising developments, and (b) the company must at some point allow the trade and sale of the founders' and early stage investors' shares, usually by going public.

Some may view such high capital profits as morally outrageous. However, one must bear in mind that the money invested at the ground stage of such innovative new ventures is definitely a high-risk investment. Statistically, most such ventures fail within the first few years, and the shares, instead of becoming worth many times their offered value, become quite worthless. However, the investor need only win 10-20% of the time in order to be a successful investor. At the same time, and more important to the development of innovative products, the venture capitalist, along with the strings he attaches to his investments and the high profits to which he aspires, provides the necessary capital to fuel the engine of innovation, including a large number of failed experiments.

Without these investors, the highly original advanced products, in for example, computer technology and biotechnology, would never be pursued. The investment philosophy of banks and pension funds is far more conservative, and is aimed at stable, predictable, and low-risk investments.

In some special cases, there may be funds available to augment those of the venture capitalist. If a national government decides that it is in the national interest to develop a strong commercial presence in a new advanced technology, then it may provide special grants, loans, or direct investment in new innovative high-risk companies.

For example, Celltech, the principal British biotechnology company, was started ten years ago as a government initiative, with an investment of some US$4 20 million. And at present, the German government is offering financial incentives to start up companies in the former DDR.

This, then, is a brief outline of the *Entrepreneurial Model* operating in a free economy, and it is the one that applies to biotechnology and other highly innovative hightech domains (such as electronics).

OPERATING A BIOTECHNOLOGY ENTERPRISE

Now let us consider some of the realities of operating a real biotechnology company in the real (not ideal) world, and the considerations that must become predominant as it solves its financing problems and begins to develop, test, and perhaps make and market products throughout the world.

Strategic Planning

So-called 'strategic planning' is the basis of determining the future course a company will follow. it is especially critical for a small company seeking niches in new and advanced technologies, where the choices of which areas/products to develop and how to maximize created value will determine whether or not the company succeeds.

Strategic planning encompasses not only the choice of products, but the means by which the development costs will be covered. Included are the timing and size of the sale of new shares of stock, as well as the planning of other mechanisms of financing, such as loans, limited partnerships, solicitations for public funds, etc. As a company grows and becomes financially stronger, it may also determine that an acquisition, merger, purchase of stock in another company, or joint venture, will be likely to enhance the company's financial strength or market position. Thus the widespread use of 'strategic' in the jargon of the financial community–one speaks of 'strategic' investments, 'strategic' acquisitions, and 'strategic' alliances, referring to any action that is expected to enhance the company's 'strategic' position with respect to its competitors.

Marketing

The selling of one's products–marketing–is an aspect of business that demands a totally different set of skills from achieving excellence in product development. Not all biotechnology companies have chosen to do their own marketing. Chiron, today one of the most successful biotechnology companies in the United States, which recently made a 'strategic' acquisition of the older but floundering Cetus Corporation, decided that at least for the first decade of its existence, it would not fall into the trap of spreading its resources too thin as would be required if it were to try to manufacture and market its own products.

Instead, the 'strategic' decision was made to concentrate on excellence in R & D, and to license its unique and valuable candidate products to seasoned marketing partners. This turned out to be a winning strategy.

On the other hand, other companies made decisions early on to 'vertically integrate', which means that they wished to become self contained companies that would do their own R & D, testing, manufacturing, and marketing. To do so, of course, requires a much greater investment than the company that sticks to research, plus the necessity to put together skilled teams to carry out all of these diverse activities. If one is successful, then one has considerable independence and earns greater profits in the end, but one also risks a difficult and expensive transitional period, during which one will have to compete with the seasoned marketing professionals.

Most biotechnology companies, even those like Genentech and Cetus, which made large early investments in developing their own manufacturing and marketing capabilities, ultimately found it in their best interests to license their best products to major international pharmaceutical or instrument companies.

After a company has gained an experienced product development group and considerable financial strength, then it may be in its best interests to consider marketing on its own. But it is usually not wise to try to do everything in the beginning, particularly if it means duplicating the manufacturing and marketing capabilities that many companies already have in place.

Rather, the successful company is the one that stresses its *unique* strengths, and these will usually lie in the originality of the new technology or products to be developed.

Nevertheless, in choosing which products to develop, it is always important to consider and understand thoroughly the market potential of any new product. This means an assessment of the size of the market–both the number of units to be sold and the price per unit--as well as the share of the market one could expect to capture. If the product is unique, badly needed, and there are no other products in sight that would compete with one's own product, then it may be possible to have, at least for a limited time, a monopoly position. While this is every entrepreneur's and general manager's dream, (a) it happens rarely, and (b) when it does, it demands a high degree of responsibility.

For example, the recombinant tissue plasminogen activator, first developed by Genentech and sold by a pharmaceutical partner, was first offered for sale at more than US$ 1200 per dose, a price that was roundly criticized for being outrageously high. When it turned out that some simpler and far less expensive drugs seemed to be just about as good in preventing second heart attacks, the sales of the high priced tPA fell off sharply, and the price was thus forced down to more reasonable levels.

While market projections for new products are understandably mostly guesswork, all investors demand such analyses before they will invest in new companies.

Unless the product has a chance of becoming a big market winner at some point, product development costs will be hard to get.

Thus, unlike commodities (e.g., food, chemicals, paper, etc.), the speciality products of advanced biotechnology, especially in the health area, must be aimed at international markets. Sometimes, in a large country, the national market may be large enough to justify investment in developing a new product, but in general, one must be concerned with the potential worldwide market for all speciality products, and how to penetrate it.

The small new company is thus usually compelled to find at least one marketing partner, or a large company with an international marketing network in place, if it expects to be competitive.

Maintaining Financial Strength

The financing strategy of start-up companies was discussed earlier. It is critically important, however, to be able to maintain and increase one's financial strength. It is especially important if a company expects to grow.

In fact, a proper growth rate for a biotechnology or high-tech company would be an increase of 50-100% per year, as measured in the operating budget or the number of employees or in 'turnover' (or 'total sales' to the accountant, which includes income from all sources and does not simply mean the revenues from product sales). A number of young companies have been posting growth rates of several hundred per cent per year.

Most biotechnology companies do not simply rely on new private placements of shares; numerous other mechanisms are used, some of which have been mentioned earlier.

For a time, Cetus was even able to survive on the interest from its record public share offering (US$ 120 million, less the 10% fee paid to the underwriters), at least while the prevailing interest rates were around 18%.

Good Management

The skills demanded of a good manager are generally very different from those of the successful entrepreneur. This is why the really clever entrepreneur should not delude himself into thinking that, just because it was possible to get a company off the ground, he is best qualified to run it.

The most successful entrepreneurs will team up with a few individuals from the beginning, among whom most of the required management skills will be represented. If these people are founders of the company, and have a personal share of stock in the company, then for sure they will have the incentive needed to do their best to see that the company succeeds.

The managers must have a vision of the future of their company and be able to make the decisions that will lead to the fulfilment of this vision. They must be decisive, and not debate alternatives endlessly before making a decision. Of course, mistakes will be made but if the leadership is seen as indecisive and weak, it will not be able to lead effectively.

Since the most effective companies are run by a management 'team' rather than by a single individual (it is virtually impossible to find all the requisite managerial qualities in any single individual), it is essential that the members of this team work well together. They should not only have complementary skills (although all should know all aspects of the business), their personalities must be compatible so that they function well on a personal, human level.

Finally, sensitivity to human reactions and needs, perhaps the most important element of good management, is absolutely essential to inspire the employees in the company and command their respect and confidence. Authoritarian management styles, still prevalent in some European and Asian companies, are generally not so effective in eliciting the highest morale and productivity from the workers. The healthiest companies are those in which the lowest workers feel comfortable in the presence of the top managers and feel free to offer criticism when it is warranted. On the other hand, the tyrannical manager who always wants to hear good news will do his company a disservice. I have personally seen a case where the heads of research groups were afraid to tell the boss that the cloning experiments were not going so well, and thus gave an over-optimistic picture of work in progress, that in turn led to bad strategic decisions. The manager who intimidates his scientists, however, must bear the responsibility for the outcome of such a situation.

SPECIAL CHALLENGES FOR COMPANIES IN TRANSITIONAL ECONOMIC REGIONS

The discussion up to this point has covered the basic economics and organizational principle of starting an entrepreneurial venture in biotechnology or other advanced technology aimed at high value-added products. To be sure, the challenge to the entrepreneur under the best of circumstances is great. In these times, when much of the industrialized world is grappling with economic stasis (the euphemism is 'recession') or even declining gross national products, it is particularly difficult for would-be entrepreneurs, no matter how exciting and lucrative their prospective portfolios, to obtain substantial financing for start-up companies.

But with the economic traditions in place, and the commercial and financial infrastructure more or less operational, it is far easier in West Europe, North America, and the Far East to be a high-tech entrepreneur than in the so called Third World (Africa, much of Asia, and Latin America) or the countries undergoing a transition from tightly controlled planned economies to a free-enterprise system. In these regions, there are additional constraints imposed on the entrepreneur. While these may differ from region to region, they certainly make entrepreneurial life more difficult. On the other hand, there may also be some advantages to new companies in these regions, if one steps back and views the world as full of new opportunities.

Lack of Entrepreneurial Tradition

In countries where there has not been much activity in the high-tech, entrepreneurial type of business described above, the aspiring entrepreneur is likely to be quite misunderstood by his compatriots, and perhaps regarded as completely crazy for suggesting starting a company that will operate at a low for five to ten years.

Those well versed in traditional business economics will tell him that his ideas are basically unsound; that he simply wants an outrageous amount of money and will be unable to offer investors any reasonable promise for a proper return on their investment.

I say this from experience, having tried to start an entrepreneurial biotechnology company in Finland, which, while it has been relatively prosperous with its mixture of capitalism and socialism, has operated more or less by the rules of traditional business economics. It has not been at all receptive to the entrepreneurial model outlined above, even when backed up by outstanding technology and prospective international markets exceeding US$ 200 million annually.

Because local thinking is slow to change, it is especially important that those who aspire to establish new ventures in such regions seek a wide base of international financial support.

Availability of Capital

Because the amount of capital available for new investment in most countries is very limited, the entrepreneur with a good business plan will simply not know where to seek his start-up financing. There are a few venture capital companies in some developing countries, most with rather limited funds; in many areas the concept of risk capital simply does not exist.

It is thus critically important for entrepreneurs in places off the beaten track to seek funds from international capital markets. An immediate constraint here is the fact that a large fraction of the countries under discussion do not have exchangeable currencies, and often impose strict rules on taking money out of the country.

Government Restrictions and Regulations

Although many national governments claim to want to encourage the development of new business in their countries, the plethora of rules and restrictions that face the entrepreneur can often be daunting. Import duties and taxes on essential items may make the capital investment required in a small developing country substantially higher than in an advanced western country.

Rules governing the establishment of corporations, the use of share capital, who may own shares (for example, many countries greatly restrict ownership of property or shares by foreigners), currency transfer and exchange, taxes, and customs regulations, can all work against the establishment of start-up companies.

Again citing my personal experiences in Finland, the policies of this country reflect the importance of the country's historical attainment of political independence and, hence, are very nationalistic. Unfortunately, Finland is a small country with small markets, and thus cannot itself support the speciality high-tech companies that require easy access to worldwide markets, as well as investment. Its nationalistic policies are at present rather out of step with current trends in Europe to reduce or eliminate all economic and trade barriers between countries. The very high costs of goods and labour in Finland, coupled with its taxation system and overvalued currency, price many Finnish products out of the reach of potential foreign buyers.

At the same time, restrictions on foreigners and foreign ownership greatly curtail investment in many key industries. The current economic depression in Finland, in part a reflection of the policies stated above, is putting increased pressure on the government to change its way of economic and business thinking.

The lesson to be learned by the newly independent European countries (e.g., the Baltic republics), and those undergoing a major economic transition, is that extending political independence to economic matters is simply wrong thinking for a small country with limited sources.

Countries that cling to the notion of economic independence and total self-sufficiency, or impose draconian measures to protect inefficient domestic industries against competition from without, are ultimately working against the best interest of both their citizens and a business community that needs all the help it can get.

Rather what is needed are liberal policies that will promote economic development by making business investments and partnerships attractive to foreign business.

Access to and Protection of Proprietary Technology

The patent laws systems of patent protection of intellectual property in many countries of the world are in great need of reform. The PCT (Patent Co-operation Treaty) is a partially successful effort to standardize and simplify international patent application and approval procedures. Unfortunately, many key countries have not signed the treaty. In countries which are not PCT members, there tends to be little legal expertise in how to obtain broad international protection of one's inventions. While patent reform is slowly taking place, the inventor today of a novel recombinant vaccine, for example, will probably have to seek counsel in a country experienced in worldwide patent coverage.

But the protection of inventions through patents is only part of the problem. It is also essential that scientists and others with novel ideas be well versed in the rules governing disclosure, so that they will not, simply by being scientists excited with their own discoveries, destroy any chance of ever receiving protection of the results of their own creativity.

Probably the classical example of how not to treat a major discovery with far-reaching practical and commercial applications, was Cesar Milstein's development of a method to produce monoclonal antibodies. The invention was disclosed to the public and no patents filed, leaving the field open to development by anyone who chose. While it may first seem desirable for such vital information to be in the public domain, free for anyone to use, in fact such a situation deters investment in product development by private capital, the reason being that without the attraction of exclusive rights to the exploitation of a discovery (and, hence, larger profits) it will be much less interesting for a company to make the requisite investment.

In other words, whether it is fair or not, the profit motive is what drives investment in the development of the exciting new products and technologies that allow us to continue to advance in all fields. Only if the rights to created novelty can be protected and secured, will there be sufficient motivation of the capital markets to pursue their commercial development.

Skilled Managers who Understand Business Economics

In countries where there is not yet a strong capitalist tradition, it may be difficult to find strongly motivated managers who really understand the business economics of the fields which they wish to enter. Business colleges and universities are beginning to train the scientist/manager/entrepreneur but this is a rare breed even in the heartland of high technology (e.g., the San Franscisco Peninsula). It is thus not only important for special training programmes to be established to train scientists in the practical aspects of biotechnology, but also programmes to educate and create a new breed of business manager, a cadre that understands the meaning of added-value, of risk, of protection of intellectual property, and of the necessity to seek broad international markets.

Scientific and Technical Manpower

However good some of the current training efforts may be, they are still grossly inadequate to meet the need for skilled scientists and technologists, who can compete with the best the West has to offer. While in the newly opened East Europe and (former) Soviet Union, there are certainly many top-level research scientists, many have been working under highly constrained conditions, and more seriously, have been more or less isolated from developments in other parts of the world.

For perhaps different reasons, the situation in much of the Third World is the same or worse, in that the level of basic training in the sciences in many countries is still several decades behind where it should be.

Economic Constraints

One possible solution for a company established in an area with inadequate skilled manpower is to import it. The problem, however, is the existence of huge salary disparities from country to country, at least in terms of international exchange rates. Therefore, one is faced with paying an imported scientist perhaps many times the salary that his indegenous colleagues would receive, or trying to recruit people to come with a substantial pay cut. As we all know, the latter simply does not happen. Needed as foreign experts may be, few will commit themselves for a long period of time where the tools of their trade and living standards they expect for their families are simply too primitive for them to accept.

Access to International Markets

While it is important that the new biotechnology companies, especially those producing high value-added speciality items, have access to wide international markets, the means to accomplish this in countries that have not been active participants in international commerce can be quite difficult. For this reason, it is very important for start-up companies to seek alliances with experienced international marketing partners from the outset.

Attracting International Partners and Investors

Although it is very important for the success of new companies to make use, where possible, of the financial strength and international experience of partners, it is not so easy to find and attract such partners. For this reason, the business idea to be pursued by the new company should have a broad international appeal. Pharmaceuticals and biologicals are generally such products, since people often (but not always) tend to contract the same or similar disease over wide areas. However, why should a large American vaccine company be interested in the prevention of schistosomiasis? Poorer countries cannot pay high prices for medicines; hence the investment in and development of such substances will not be undertaken by companies when they can gain much higher returns with medicines aimed at the diseases of the wealthier countries.

This is where the entrepreneurs must also be inventive. It is often the case that a technological approach to producing a new substance can be applied to more than one objective. And it can also be the case that while the new company may not be able to interest a western company in a specific product for a specific indigenous disease, it may have other resources or technology of direct interest to a major company. Thus it is not ruled out that a large multinational would make a strategic investment in a company in a developing country in order to gain access to particular biological resources, or future markets.

CONCLUSIONS

What I have intended as an introductory lesson in entrepreneurial business economics will no doubt leave many of my listeners unsatisfied. I expect that I have raised more questions and problems than I have answered, and I have certainly not given every prospective entrepreneur a recipe to follow that will guarantee success the first time.

In fact, I may have presented so many sources of difficulty (without solutions) that face the entrepreneur in a country in a state of economic change, that even the most adventurous will be daunted. I hope not.

Rather I hope that I have successfully described the very exciting opportunities that await those with confidence in their creativity, intellect, and perhaps, living by their wits, who wish to take an active part in the economic revolutions that are now sweeping many parts of the world, in part made possible by the revolutionary technologies that excite all of us.

Personally, I like a good challenge. Attacking difficult problems is fun, and very satisfying when one succeeds. And even if one does not, one always learns some very useful lessons for the next time around. As in science and engineering, there is still a great deal of room in applied economics for the original, creative approach to new business, and to mobilizing the vast scientific and technical resources at our disposal.

I truly hope that many of you will start to think of every difficult situation as an opportunity and a challenge, that simply awaits an original solution. If you do, then you are on your way to becoming a successful entrepreneur.

5

LEGAL ASPECTS AND INTERNATIONAL DEPOSITORY AUTHORITIES

C.B. MORRIS
ECACC, PHLS-CAMR, Porton Down, Salisbury, England

A subject of increasing importance to the bench scientist is the understanding of patent law. This knowledge has become necessary owing to the fact that each year scientific discoveries in the biological fields yield potentially exploitable products. The areas encompassed by biotechnology are no exception, and today's market value of healthcare and agricultural products alone runs into billions of dollars.

The essential points of the patent procedure are outlined to provide the reader with a practical understanding of the subject.

The initial question raised is why should I patent my discovery? The answer lies in comparing the alternative, that is maintaining total secrecy. An inventor choosing this option could, in theory, commercially exploit his invention for as long as he can prevent others from copying the invention. To overcome this inevitable situation, the discovery of a new invention would have to be totally suppressed and thus many valuable discoveries would be lost.

To encourage openness of discovery, the inventor is granted a legally enforceable right to prevent others from exploiting their invention. This protection is, however, for a fixed time period. In order for an invention to be accepted for patent protection it must pass several criteria. These are:

Novelty–the invention must be new or have 'absolute novelty'. This precludes the possibility of prior knowledge, in practice meaning that details of the invention must not already be known, i.e., have been communicated verbally or written. Canada and the USA however allow publications to appear two or one years, respectively, prior to a patent application.

Invention–the invention must not be obvious from existing knowledge and must show evidence of an 'inventive step'.

Utility–the invention must have a practical use. Most countries, excluding the USA, insist that it must have an industrial application.

Disclosure–the invention's description must provide sufficient detail, such that a skilled person can reproduce it. In the case of inventions involving micro-organisms, which includes animal, human, and plant cells, most countries require that a deposit of the organism is made in a recognized culture collection.

This is a unique feature of biotechnological patents arising because a written description will not usually provide sufficient detail for total disclosure.

Collections recognized for deposit of patents are termed International Depository Authorities (IDAs) of which more details are given below.

Biotechnological patents fall into four main groups, namely:

Products–which covers new materials or entities. Included in this group are living organisms, e.g., bacteria, fungi; parts of an organism, e.g., animal and plant cell lines; products, e.g., enzymes; plasmids and DNA.

Compositions–mixtures of organisms or substances.

Processes–methods for manufacture or purification.

Methods–including processing materials, e.g., raw materials industrial by -products, quality control testing, and diagnostic methods.

As mentioned previously, disclosure of a patent must include its deposit in an IDA. To overcome the problem of multiple deposits in different countries, the ***Budapest Treaty on the International Recognition of the Deposit of Micro-organisms for the Purposes of Patent Procedure*** was created under the auspices of the World Intellectual Property Organisation (WIPO, 1980, 1988).

By agreement of the parties to the treaty, a single deposit made with an IDA in any of the participating countries will provide immediate protection. However, if patent status is ultimately sought for several countries, this application must be converted to separate national and regional applications. The term regional has been used in this case to cover the European Patent Convention (EPC), which consists of 14 participating European countries.

In the same way that a single application can be lodged under the terms of the Budapest Treaty, it is also possible to make a single application for countries through the European Patent Office (EPO) in Germany.* In this instance, the terms of acceptance vary from those of the Budapest Treaty and are discussed further under the requirements for disclosure and release of samples.

IDA'S

To become an IDA, a culture collection must fulfil several obligations the most critical being that it resides in a member country and can provide suitable storage facilities for that organism or those organisms for at least 30 years. An IDA is not obliged to accept the full range of organisms but may increase its range by application to WIPO, as in the case of the author's laboratory which has recently extended its patent acceptance to include viruses, recombinant DNA, and plant cultures, in addition to human and animal cell lines.

A useful source of information concerning the patent status of European collections is the Information Centre for European Culture Collections (ICECC).**

Each IDA sets its own terms of acceptance, i.e., minimum number of samples, number of organisms per sample, viability and quality control criteria, etc. Additionally, the IDA will require notice of a deposit, to ascertain its suitability for that collection, and for application forms to be completed.

It should be appreciated that an application cannot be filed until the patent office receives the succession or deposit number given to

* *European Patent Office, Erhardstrasse 27, W-8000 Munchen 2, Germany.*

** *ICECC, Mascheroder Weg 1b, D3300 Braunschweig, Germany.*

the deposit by the IDA. This becomes the filing or 'priority' date and will establish the priority of that application over any others made.

The deposit number is normally given on the day the sample is received. However, final acceptance of a deposit will depend on its passing the IDA's quality control tests. For this reason it is essential to allow sufficient time for the IDA to not only receive the sample, i.e., to obtain import permits and customs clearance, but to undertake their tests. This can require up to two months, especially when special growth media have to be prepared. This can be expedited if the depositor always provides any special growth supplements with the deposit.

Of practical consideration is the eventuality that a deposit fails its quality control tests, e.g., viability or contamination, but that the depositor has already filed their application. Two possibilities follow: either a new sample is deposited, which will receive a new, i.e., later, deposit number or the IDA attempts to restock from the original deposit. If the latter is successful, the original deposit number is retained.

From the author's experience with animal cell lines, contamination with mycoplasma is common. This has frequently led to the problem of eradication, in order to retain the original deposit number. A depositor is therefore strongly advised to undertake a full quality control of their organism, prior to depositing a sample.

As an illustration of an IDA's acceptance and quality control requirements those of the ECACC for animal cell lines are listed below:

DEPOSIT REQUIREMENTS

- minimum of 48 hours notice of arrival
- minimum of 12 frozen ampoules (all frozen at the same time from the same batch of cells)
- minimum of 4×10^6 cells ampoule
- maximum ampoule size 23 mm dia., 50 mm height
- all ampoules to be clearly marked with a permanent ink
- special growth factors to be supplied by the depositor
- pathogenicity status and handling conditions to be declared prior to receipt.

QUALITY CONTROL

- after resuscitation, a viable and total cell count is made
- cells are cultured for one week free from all antibiotics
- two Hoechst stain preparations on Vero indicator cells examined for myoplasma
- sample taken for bacteria and fungi testing
- sample taken for mycoplasma testing in broths and on agar (minimum of five weeks required to complete the tests).

RELEASE OF SAMPLES

Part of the concept of intellectual property under patent law is its availabilty to third parties. Currently there is no single agreement on the mode of release of samples, but rather they generally fall into three possible procedures depending on which country the application is lodged in.

USA–because patent applications are not published prior to granting a patent, its existence remains secret until this time. This confers the advantage that the depositor need not release his organism until he has full legal protection.

Japan and The Netherlands–in this case a list of patent applications is published 18 months after the filing date and also on its acceptance. Request for release of a deposit can only be made after the patent is granted. Additionally in Japan, the recipient must not pass on the organism to third parties.

European Patent Convention–deposits made through EPO can be released at the date of application, i.e., priority date. This has the obvious disadvantage of release before full protection is granted. The reasoning behind this approach is that the EPC considers the organism an integral part of disclosure, and should be made available on request. Recent amendments to the original ruling allow the depositor to release their organism to a chosen independent expert acting for the requesting party. A list of experts is held by the EPO, each having agreed not to pass on cultures to other parties.

The whole process from filing a patent application to acceptance (i.e., grant) will normally take between 2 to 3 years. The aspects of suitability, examination, and appeals against the application, can only be dealt with by a patent office specializing in intellectual property. Advice on this matter can be obtained from the EPO or your national patent office.

Additional useful information can be obtained from the references (Crespi, 1988; Bousfield, 1990; Crawley).

REFERENCES

Bousfield, I.J. 1990. Patent protection for biotechnological inventions. In *Living Resources for Biotechnology: Animal Cells.* Cambridge University Press, England. pp. 101-50.

Crawley, P. *Patenting in Biotechnology for Non-specialists: An Introduction with Sources of Further Information.* From Bio-industry Association, 1 Queen Annes Gate, London SW1IH 9BT, England.

Crespi, R.S. 1988. *A Basic Guide to Patenting in Biotechnology.* Cambridge University Press, England.

World Intellectual Property Organisation. 1980. Records of the Budapest Diplomatic Conf. Conclusion of a Treaty on the International Recognition of the Deposit of Micro-organisms for the Purposes of Patent Procedure. World Intellectual Property Organisation, Geneva, Switzerland. No. 332 (E), p. 119.

World Intellectual Property Organisation. 1988. *Guide to Deposit of Micro-organisms for the Purposes of Patent Procedures.* WIPO , Geneva, Switzerland.

6

EUROPEAN CULTURE COLLECTIONS: FUNCTIONS, SERVICES AND INTERACTIONS

K.A. MALIK

DSM—German Collection of Micro-organisms and Cell Cultures, Brauschweig, Germany

In microbiology and biotechnology, there is an ever-growing need for a constant and stable source of supply of micro-organisms, which is usually managed in a well organized manner by culture collections. The common activities of collections which are described in this article are the established capabilities and competence of various leading culture collections worldwide, which with very few exceptions are mostly located in Europe. The world's first known culture collection was also established in Europe some hundred years ago by Frantisek Kral (1846-1911) in Prague, Czechoslovakia.

European culture collections began as a resource primarily for micro-organisms but now many of these have a broader role as world gene banks, resulting from new developments in biotechnology. These culture collections have proved their worth by preserving important and diverse groups of prokaryotic and eukaryotic strains and offer a broad range of services for biology, biotechnology, and industry. However, many collections especially in the eastern part of Europe are poorly organized and inadequately funded. As a result, many regionally useful species of micro-organisms and germ plasm, found chiefly in such areas, may become extinct. Therefore, a stable national and international germ plasm base is required for maintaining our genetic resources once we have acquired and assembled them, be they aquacultures, microbes, plants, seeds, or animal cells.

The service culture collections act as depositories for all kinds of micro-organisms that are of past, present or potential importance and thus function as resource and information centres for the general support of microbiology and biotechnology. For service collections, it is of fundamental importance that they adopt a selection of maintenance methods that produce maximum survival levels and guarantee strain stability.

The major activities common to all European service culture collections are that they generate a quantity of essential information which is now systematically recorded using computers. Strain data stored in a computer allow the search and retrieval of scientific information and establishment of a data base that may be used subsequently for computer identification and for the preparation of a catalogue of strains. An up-to-data catalogue is an essential requirement for the effective functioning of a service collection, and the transfer of catalogue data to computer allows a ready updating for new editions.

Microbiology and biotechnology not only need viable and stable cultures but also organisms that are properly named and documented. Organisms with defined properties can be used as reference and production strains. Thus the information about the systematic name of an organism is the key to a wealth of useful ecological, medical, genetical, and biochemical data.

Microbial strain identification and data management are mostly included in the functions and services of European culture collections and are offered as service to facilitate the identification of and search for important microbiological data. Based on their collection-related research programmes the major European service culture collections generate essential and useful information about their holdings which is not recorded systematically in their catalogues using computers.

Strain data stored on computer allow the search and retrieval of scientific information as give information about the usefulness of a strain. The World Data Centre* for micro-organisms in Japan houses a master copy of the World Directory of Collections of Micro-organisms and provides information on collections and their holdings

* WDC, Riken, 2-1 Hirosawa, Wako, Saitama 351-01, Japan.

worldwide, especially to the Third World. The WDC makes several databases available through the European Microbial Strain Data Network.** The MSDN provides a Directory to sources of information on micro-organisms and cultured cells.

In order to have improved access to the wealth of information available in European culture collections and for publicity, various information networks and databases such as the Microbial Culture Information Service (MiCIS), the Microbial Information Network Europe (MINE) and the Information Centre for European Culture Collections (ICECC, based at the DSM) have been founded. The European Community (EC) strongly supports these activities in Europe.

All well established culture collections play several other important roles in addition to the careful deposition, documentation, maintenance, preservation, and distribution of useful, representative and type strains. The main interest of some bacterial and fungal service culture collections in Europe and their type of holdings and services offered to microbiology and biotechnology has already been summarized in other articles (see selected references). Such collections are a world resource whose significance may only be recognized in the light of future scientific discoveries. The strains collectively maintained between the collections also provide a tremendous genetic resource of biotechnological importance for the present, and could prove to be of immense significance for the future.

All preserve cultures not only should retain maximum viability over prolonged periods of storage but they must remain unchanged in their properties. As often observed, during inadequate maintenance and preservation, strain drift occurs and industrial production strains fail to show consistency of performance. Therefore, maintenance and preservation of a full range of isolates is an enormous task and even the largest collections are generally not able to build up large holdings of strains of single species, genera or groups. Therefore, major European culture collections build up gene banks that they judge valuable or that would be useful in the future, e.g., strains including type strains, strains deposited for their special properties, assay strains, genetically engineered strains, industrial strains, plasmid-bearing strains, and others. Once acquired, such strains are authenticated, documented, and subjected to preservation using the most appropriate

** MSDN, Institute of Biotechnology, Cambridge University, Cambridge CB3 0XJ, England.

methods in order to function as a continuous source of supply for viable and stable strains (see Fig. 6.1 and 6.2)

Fig.6.1: Liquid nitrogen storage of stock cultures at the DSM.

Fig.6.2: Storage of freeze-dried micro-organisms in ampoules at the DSM.

The preservation of micro-organisms and animal or plant cells is a highly specialized activity calling for experienced back-up and often very specialized equipment. It is not possible for the collections to optimize the preservation of each culture and put forward a universal method of preservation, since as often observed, there is a great variation in susceptibility to the various preservation methods and usually the method suitable for one group, genera or species might not be suitable for another. With inadequate preservation, some micro-organisms could also undergo changes in physiology, biochemistry, pathogenicity, or other valuable characteristics. Thus the collections in Europe that are responsible for maintaining a broad spectrum of isolates over long periods in a viable and stable state mainly have adequate equipment, personnel, and financial resources to operate a variety of modern preservation methods.

Collections serve as repositories for organisms cited or used in scientific publications so that the micro-organisms can be used for reference work and the original work can be repeated or confirmed.

Culture collections carry out research programmes that complement their interests and have the advantage of access to large numbers of authenticated micro-organisms available for study.

The majority of micro-organisms in a culture collection are often not fully characterized and provide the possibility of being explored and screened for their full research and industrial potential. The expertise developed at the collections enables the collection staff to give advice on taxonomy, identification, isolation, culturing, maintenance, preservation, supply, deposition, and other collection-related matters.

The interaction and collaboration between culture collections through national, regional, and international networks is now rapidly increasing. The World Federation of Culture Collections (WFCC) promotes and co-ordinates such activities at international level and the European Culture Collections' Organisation (ECCO) in Europe.

The ECCO, established in 1981, now has about 43 member collections and it provides a chance for the curators to meet annually to discuss collection-related problems in Europe. Many ECCO member collections function as International Depository Authorities (IDAs) and are able to provide necessary information to patent processes and applications.

During the Tenth Meeting of ECCO in Valencia, Spain in June 1991 it was proposed that ECCO should join the European Federation for Culture Collections. ECCO hopes that the culture collection community has an essential role to play in the isolation, preservation, and conservation of the microbial biodiversity, and has the necessary expertise available for this.

ECCO is a adhering member of the International World Federation of Culture Collections and promotes the aims of the Federation. For more details on information, services, and stock cultures in European culture collections, and on ECCO, the reader is referred to the ICECC, News Letter.*

The usefulness of culture collections throughout the world and their fundamental role in microbiology has often been projected in the International Culture Collection conferences and by the world network of Microbial Resources Centres (MIRCEN). The leading culture collections, particularly from Europe, are very actively involved in collaborative research and training of personnel from the developing countries and from the less developed parts of Europe.

These represent a feeding channel to the MIRCENs, an international network, and a constant source of supply for their working tools, micro-organisms and cell cultures. Today, there are 19 MIRCEN centres worldwide which participate in a global collaborative network, particularly in the harnessing of the beneficial applications of the Microbial world for human progress.

On the occasion of a European MIRCEN Biotechnology Symposium at Ljubljana, Yugoslavia, in 1988, it was decided to establish a European Network of Microbial Resources Centres around the existing European culture collections in order to expand collection related functions, services, and research activities.

It was also decided to establish a Biotechnological Information Exchange System (BITES) that complements the work of the European MIRCENs.

In short, in culture collections, substantial efforts are being made in Europe to establish co-operation in the fields of acquisition, study, distribution, training, and proper usage of microbial resources available through culture collections and allied institutions.

---

* Acting Secretary ECCO, ICECC, Mascheroder Weg 1b, D 3300 Braunshweig, Germany.

REFERENCES

Allsop, D., Hawksworth, D.L. and Platt, R. 1989. The CAB International Mycological Institute culture Information Service (MiCIS) and Microbial Information Network Europe (MINE). *Int. Biodeter.*, 25, 169-74.

Da Silva, E.J., Burgers, A.C.J. and Olembo, R.J. 1977. UNESCO, UNEP and the international community of culture collections. In F. Fernandes and R.C. Pereira (eds), *Proc. Third Int. Conf. Culture Collections, 14-19 March 1977.* University of Bombay, India. pp. 107-20.

Hawksworth, D.L. 1985. Fungus culture collections as a biotechnological resource. In G.E. Russel (ed), *Biotechnology and Genetic Engineering Reviews.* Intercept, Ponteland, Newcastle Upon Tyne, England. Vol. 3, pp. 417-40.

Hill, L.R. and Krichevsky, M.I. 1986. International strain data network. MIRCEN J., 2, 341-47.

Kirsop, B.E. 1987. Microbial culture collections: repositories of microbial gene polls. *Nature and Resources (UNESCO, Paris)*, 23 (2), 2-9.

Kirsop, B.E. and Doyle, A. (eds). 1991. *Maintenance of Micro-organisms and Cultured Cells.* Academic Press, London.

Malik, K.A. 1987. The role of culture collection in the stability and preservation of micro-organisms. In J. Amen and P. Tesson (eds), Ï2eme Colloque, la Section de Microbiologie Industrielle et de Biotechnologie de la SFM, Societe Francaise de Microbiologie, Stabilite et Conservation des Microorganismes. pp. 118-50.

Malik, K.A. and Claus, D. 1987. Bacterial culture collections: their importance to biotechnology and microbiology. In G.E. Russel (ed), *Biotechnology and Genetic Engineering Reviews.* Intercept, Ferndown, Dorset, England. vol. 5, pp. 137-97.

Malik, K.A. 1988. Preservation of biotechnologically important micro-organisms in culture collections. In A. Blazej and J. Zamek (eds), Proc. Interbiotech 87, Enzyme Technologies, Elsevier Science Publishers, Amsterdam, Holland. *Prog. Biotechnol.*, 4, 145-86.

Malik, K.A. and Hoffmann, P. 1989. Preservation and storage of biotechnologically important micro-organisms. *Chimicaoggi*, June, 61-66.

McGowan, V. and Skerman, V.B.D. (eds). 1982. *World Directory of Collections of Cultures of Micro-organisms.* World Data Centre for Microorganisms, Brisbane, Australia. 2nd edn.

MIRCEN News 9contact E.J. Da Silva, Div. Scientific Research and Higher Education, UNESCO, F-75007 Paris, France).

Porter, J.R. 1976. The world view of culture collections. In R.R. Colwell (ed), Proc. American-Type Culture Collection 50th Anniversary Symp., *The Role of Culture Collection in the Era of Molecular Biology.* American Society for Microbiology, Washington, DC, USA. PP. 62-72.

Takishima, T. *et. al.*, (eds). 1989. *Guide to World Data Centres on Micro-organisms*, with list of culture collections worldwide. RIKEN, Saitama Prefecture, Tokyo, Japan.

7

SETTING UP AND MAINTAINING A PLANT CELL CULTURE COLLECTION

H.M. SCHUMACHER
DSM-German Collection of Micro-organisms and Cell Cultures, Braunschweig, Germany

Summary. *In vitro* cultured plant material today serves as a tool for many purposes in agriculture biotechnology, and fundamental research. This gives rise to the need to preserve material which has turned out to be of special scientific or practical importance. In general, one can distinguish between *in vitro* cultured shoots or plantlets, and tissue or cell cultures. *In vitro* cultured shoots or plantlets are normally maintained on solid growth restrictive nutrient media at reduced temperature. Depending on the plant, one can achieve subculture periods from several months to several years. Genetic stability is assured when axillary buds are used for propagation.

A comparable degree of genetic stability is also provided by *in vitro* cultured plant organs, like shoot- or root-cultures. Growth retardation can be achieved best by culture on solid media under reduced temperatures.

The situation with undifferentiated callus or suspension cultures is more problematic. Cultures are established from fresh plant material (leaves, stems, roots, or seedlings). Tumour-like callus growth is achieved under the influence of certain plant hormone regimes. These cultures exhibit a high degree of phenotypic and genotypic instability.

which is used to select out strains with certain desired metabolic potentials. Failure to maintain the selection pressure can result in the loss of this potential. Unfortunately, no general method is available for preservation or characterization.

For most applications, suspension cultures are used. Because of their vigorous growth, it is almost too laborious and too expensive to maintain cell cultures as suspensions. Most laboratories use callus cultures for long-term maintenance. Various slow growth conditions have been established, like reduced temperature or osmotic stress. Callus cultures can also be preserved for more than one year under a mineral oil layer.

The only available long-term preservation method is freezing and storage at liquid nitrogen temperatures. Special problems are caused by the high water content of plant cells.

Different cell cultures show different behaviour towards all the preservation methods. Moreover, all preservation methods even cryopreservation imply the danger of undesirable selection processes. Careful characterization of cultures before and after preservation is therefore necessary.

At present, undifferentiated cell cultures are most often used as experimental tools for fundamental research but in certain cases also for large-scale production. Since 1987, the DSM (German Collection of Micro-organisms and Cell Cultures) has run a collection of undifferentiated cell cultures. It presently consists of more than 1000 cultures from 44 different plant families. With careful use of suitable preservation and maintenance methods, in vitro cultured plant cells may provide a novel approach to overcoming present problems of germplasm conservation with plants.

THE NEED FOR CONSERVATION

The need to conserve plant genetic resources is evident and the reasons are numerous. It is estimated that more than 350,000 plant species exist at present on Earth. Only a small number of these plants have been investigated for their secondary metabolite content. A lot of plants might contain useful compounds or exhibit certain useful traits or genes still unknown (Baerheim-Svendsen, 1984). In many cases, wild type genes have successfully been used to improve important crop plant varieties. Semi dwarf mutants, for example, acquired considerable importance as rice and wheat growers. When new diseases appear, genes from wild-type plants or old varieties are needed immediately to maintain even the present standard of production. When the cereal leaf beetle reached Michigan in the 1960s, genes from old wheat varieties obtained from Russia in 1927 were used to breed resistant cultivars (Elgin and Miller, 1989).

Normally, seeds are the ideal tool for plant preservation, created by nature itself. Seeds can be dried and stored in genebanks at reduced temperature. Nevertheless, in many cases other methods are needed. The necessity for seed propagation when a certain degree of viability loss is reached sets limits in the case of plants that are difficult to grow (date palm). Many other plants produce short-lived so-called 'recalcitrant' seeds which cannot be stored at all. With vegetatively propagated plants, the traits of certain cultivars are not stable during propagation via seeds. This is also true for new plant varieties obtained from *in vitro* techniques for plant improvement as somaclonal variation of somatic hybridization. In these cases, cell and tissue cultures can provide material for the preservation of plant genes.

The increasing importance of undifferentiated cell cultures for biotechnology as well as for fundamental research is in the same way dependent on the preservation of resources. When a cell culture exhibits a specific selected trait or its embryogenic potential is lost, there is no alternative to the preservation of the cell culture.

The above-mentioned arguments demonstrate that alongside sanctuaries, botanical gardens, field- and seed-genebanks, cell culture collections can also make an important contribution to the preservation of plant genetic resources (Towill and Briggs, 1989).

STRATEGIES FOR CONSERVATION

Different kinds of *in vitro* cultured plant material and different conservation strategies exist. The most important cell and tissue culture types for conservation are undifferentiated callus or suspension cultures on the one hand and meristem or shoot-tip cultures on the other hand. To meet the needs of agriculture, material should be conserved from which whole plants can be obtained. This is possible from zygotic embryos as well as from embryogenic callus, gametophyte, or pollen culture and meristem or shoot-tip culture. In most cases where zygotic embryos are available, conservation can also be executed at seed level. Pollen are very suitable tools for preservation but can only be used for the conservation of part of the genome. The conservation of embryogenic calli implies the risk of genetic instability from somaclonal variation. The most suitable *in vitro* materials for agriculture are undoubtedly meristem or shoot-tip cultures. They offer the advantages of *in vitro* conservation and also provide the necessary genetic stability (Towill and Briggs, 1989).

When cell culture material for application in biotechnology or fundamental research is to be preserved, there is no alternative to the preservation of the cell culture itself. Preservation of the original plant instead of the cell culture is not satisfactory, because newly established cell cultures of the same plant do not always exhibit the same metabolic characteristics. Furthermore, very often selection procedures have been applied to establish a cell culture line with a certain metabolic potential. The specific characteristics of cell culture lines may be due to mutations of the original plant genome as well as to phenotypic differences. In both cases, only the preservation of the callus or suspension culture preserves this specific potential.

The two major *in vitro* culture types, shoot-tip and callus/ suspension cultures, can be maintained in different ways. In contrast to pollen or embryos, both culture types can be maintained in a living state by continuous subculturing. This is normally carried out under optimum growth conditions. It normally leads to vigorous growth and the need for frequent transfers to fresh medium. The method is therefore laborious and extensive. But frequent subculturing also enhances the danger of losing cultures by infection. While shoot-tip cultures provide genetic stability, in the case of callus or suspension cultures strong growth of a culture also enhances the risk of genetic or epigenetic changes (Evans *et al.*, 1987).

An alternative strategy is maintenance under slow-growth conditions. This is applicable to both culture types. It lowers costs and expenditure. The risk of infection is also reduced but still exists. The risk of genetic instability is decreased because of the reduced number of cell divisions per time interval. On the other hand, slow growth conditions might expose the cultures to a certain undesired selection pressure.

The most suitable method is cryopreservation. Freezing cell or tissue cultures to liquid nitrogen temperatures means the total suspension of vital functions. It, therefore, provides complete genetic stability also for callus or suspension cultures.

Once frozen, cells cultures are no longer in danger of being infected and can be stored cheaply and safely. Only preservation by dessication or freeze-drying might be even more advantageous.

Unfortunately, the high water content of cell cultures sets certain limits on the applicability of deep freezing. Since the most severe damage during freezing procedures is caused by ice crystal formation in the cytoplasms or the vacuole, cells with very low water content are more amenable to the application of freeze preservation.

TECHNIQUES FOR PRESERVATION OF CELL AND TISSUE CULTURES

Shoot-tip Cultures

As mentioned already, shoot-tip cultures are mainly used to preserve material for agricultural use. Cultures are initiated from excised shoot tips. This technique offers the possibility of eradicating viruses or other infective diseases. In the case of potatoes, nodal segments from germinated tubers or stem segments containing at least one axillary bud are the starting material. After initiation, the cultures are normally grown as plantlets in small glass vessels. The medium can be solidified with agar but the cultures can also be grown in liquid media fixed in a carrier system like rockwool. Ethylene production might influence growth when vessels are tightly closed. In some cases, growth conditions and conditions for culture establishment differ between different cultivars. When plant growth is to be re-established from *in vitro* culture, rooting and a certain adaptation phase to normal environmental growth conditions are necessary. In these cases, different cultivars may have different requirements (Mix, 1985).

When cultures are grown under optimum conditions, subculture periods range from several weeks to several months. Subculturing itself means only transferring the plantlets to fresh medium. But in most cases where *in vitro* culture is practically applied for germplasm preservation, cultures are grown under slow growth conditions. The periods for subculturing can be prolonged to several years. A period of four years might be reached for apple trees. Slow growth is obtained either by growth at reduced temperatures or with the application of growth retardants like absisic acid, N, N-dimethylsuccinamic acid or B9 or the application of both methods. When cultures are maintained under slow-growth conditions, subculturing in most cases includes a recovery phase of the cultures under normal growth conditions.

Regrown to a certain level under normal conditions, the cultures are then exposed to slow-growth conditions again (Mix, 1985).

The long time between subculturing periods reduces the danger of infections during the transfer of cultures to fresh medium. On the other hand, the long period enhances the danger that vessel caps or stoppers become loose. Additionally slow-growth conditions, also reducing microbial growth, make infections more difficult to detect.

Although freeze preservation of whole plantlets is impossible, it has already been successfully applied to shoot tips. Two different methods have been developed for the cryopreservation of isolated shoot tips. Generally, damage from ice crystal formation in the cell interior must be avoided. With shoot tips, this can be done by slow freezing as well as very rapid freezing. In the case of rapid freezing with freezing rates of several hundred degrees centigrade per minute, the temperature range where ice crystals can form should be overshot so rapidly that freezing occurs in an amorphous state. In practice, it has been carried out simply by dipping shoot tips fixed on needles directly into liquid nitrogen. The method has been applied for potato meristems (Henshaw *et al.*, 1985).

Slow freezing seems to be the more efficient and more generally applicable method. Here, intracellular ice formation should be avoided in a multistep procedure. The water content of the excised shoot tips is reduced by preculture in high osmatic media. Freeze protection is achieved by adding cryoprotectants to the preculture medium or to the cryosolution. Further reduction of the water content is achieved when the shoot tips are cooled down with slow freezing rates 0.5°C/min-2°C/min). Ice formation in the external medium increases the solute concentration in the outer medium which leads to a further water loss of the cells. Slow as well as rapid forzen shoot tips are normally thawed with fast thawing rates (Kartha, 1985).

A necessary precondition for applying cryopreservation is the existence of suitable regeneration methods for the thawed shoot tips. While genetic stability is assured during the storage period, selection processes might occur during the freezing procedure. It is also important that thawed shoot tips might form shoots via callusing. Those derived from callus shoots might be genetically different from the original shot tip.

Cryopreservation has been successfully carried out with different species, but it has still not been applied in practical genebanking.

Callus and Suspension Cultures

Undifferentiated call cultures are obtained when sterilized plant material is exposed to plant-hormone-containing media. Plant hormones induce callus growth. These calli are normally maintained on agar-solidified media. In 1987, the DSM started to run a collection of undifferentiated callus cultures. The basis for this collection were donations by Zenk, from Munich and the Nattermann Company in Cologne. At present the cultures are maintained in optimum growth conditions by continuous subculturing.

Subculture periods for fast-growing calli are 3-4 weeks. Some calli release toxic compounds into the medium and have to be transferred to fresh medium before the nutrients are spent or the medium has become too dry. Normally suspension cultures are the application form of undifferentiated cell cultures. They are derived from calli. Suspensions grow more homogeneous but also much faster than calli. They have to be transferred to fresh medium every week. Moreover, their growth also needs expensive equipment because they have to be shaken continuously. Preservation of undifferentiated cell cultures in a living state therefore normally uses calli, especially since slow growth methods have been found to be rather unsuccessful, and cryopreservation methods are difficult to establish.

The major disadvantage of maintaining living cultures is the high degree of genetic aberrations occurring in a callus culture. Callus cultures consist of a highly heterogeneous cell material, and the detection of many chromosomal aberrations as well as of somaclonal variation among regenerated plants revealed that genetic differences and changes are the basis for this heterogeneity (D'Amato,1978). Additionally, epigenetic changes take place as can be seen from selected cell lines that fall back into the original state when selection pressure is not maintained. The problem of identical preservation of living undifferentiated cell cultures can at present only be solved by careful subculturing carried only be solved by careful subculturing carried out by skilled and responsible technical personnel.

As mentioned above, in contrast to shoot-tip cultures, the application of slow-growth methods was found to be rather unsuccessful with callus cultures. Of more than 600 cultures tested, only about 40 survived for more than 3 months at 4°C.

Even more developed than with shoot tips is the state of cryopreservation in the case of undifferentiated cell cultures.

Unfortunately, in most cases calli cannot be frozen to liquid nitrogen temperatures. Since most cultures are grown as calli, a suspension culture has to be established before freeze preservation can be carried out.

Cryopreservation with suspension cultures meets the same problems as with shoot tips. Using cells with a low water content is generally important. With many suspension cultures, this can be achieved by taking cells from the early logarithmic growth phase. Pregrowth in highly osmotic media leads to a further reduction of the water content. In the same way as with shoot tips, slow freezing with a controlled cooling rate is applied. For freezing and storage of the cells, commercially available plastic cryo vials are used.

Thawing of the cells is normally carried out simply by pouring cryo vials into a 40°C water bath. Recovery growth has received more attention. It has been shown that in most cases, recovery growth on semi-solid media produces the best results. Neither washing of the cells nor regrowth in liquid media turned out to have any advantage.

As with shoot tips, storage at liquid nitrogen temperatures itself is ideal and does not imply any severe danger. But the freeze preservation procedure and recovery growth might induce changes of the culture or give an undesirable selection pressure. Cell cultures should therefore be carefully characterized before and after the cryopreservation procedure. Cultures tested so far have turned out to be unchanged by the procedure both qualitatively and quantitatively (Withers, 1985).

The fact that many selected high producing cell lines for large-scale production purposes have lost viability during cryopreservation procedures has been a significant drawback. This might be due to a self-intoxication of the cells when secondary metabolites are over-concentrated in the vacuole as a consequence of dehydration. Nevertheless, improved methods will increase the importance and routine application of cryopreservation of cell culture collections.

ORGANIZATION OF A CELL CULTURE COLLECTION

Normally, a cell culture collection has not only the task of preserving material but also of supplying the material to customers and clients. As long as the cultures are maintained in a growing state, material to send off to clients can be taken from the routinely grown cultures. When a certain preservation method is applied, it has been suggested that the material be divided into a base collection and a collection used for supply and propagation.

When material has to be maintained in a living state, different precautions have to be taken to avoid risks of losing cultures. New material has to be carefully investigated for infections before it is included in the collection. At our collection, two sets of the whole collection are stored in different places. This prevents occasionally occurring infections from spreading over all samples from the same culture. The two culture sets are maintained by different technicians. This enables us to monitor the cultures of the different sets independently for changes or alterations in the growth characteristic. It also has the consequence that for culture transfer of the different sets, normally different medium preparations are used, so avoiding culture loss by deficiently prepared media.

Another problem that should not be underestimated is culture loss by mistaken marking of cultures. In the case of undifferentiated cell cultures as well as with shoot cultures, the plant species or cultivar from which the culture has been derived cannot be determined from culture material itself. In the case of non-embryogenic undifferentiated cultures, the situation is even worse. Since chromosomal aberrations prevent even the application of investigation of the karyotype, no method exists which allows identification of the plant from which a culture has been derived. Once mixed or incorrectly marked, cultures are lost for any further use. Therefore, methods for the characterization of cell cultures should be developed to control the true-to-type preservation of cultures during a certain freezing process. Characterization might be carried out by measuring secondary metabolite 'fingerprints' or by genetic fingerprints obtained from RFLP methods.

At present owing to inadequate preservation methodology and experience with long-term preservation of *in vitro* cultured plant material, conservation of plant genetic resources by *in vitro* methods still implies a small risk of losing genes or metabolic potentials. The best way to overcome these risks is simply to gain more experience of running a cell culture collection.

REFERENCES

Baerheim-Svendsen, A. 1984. Biogene Arzneistoffe - heute noch oder heute wieder. In F.C. Czygan (ed), *Biogene Arzneistoffe.* Vieweg, Braunschweig. pp. 27-42.

D' Amato, F. 1978. Chromosome number variation in cultured cells and regenerated plants. In T.A. Thorpe (ed), *Frontiers of Plant Tissue Culture.* IAPTC/University of Calgary, Canada. pp. 287-95.

Elgin, J.A. and Miller, P.A. 1989. Enhancement of plant germplasm. In L. Knutson and A.K. Stoner (eds), *Biotic Diversity and Germplasm Preservation–Global Imperatives.* Kluwer Academic Publishers, Dordrecht, Boston, London. pp. 311-22.

Evans, D.A., Sharp, W.R. and Bravo, J.E. 1987. Plant somaclonal variation and mutagenesis. In Nestec Ltd. (ed), *Nestle Research News.* Nestec Ltd., Veyvey, Switzerland. pp. 63-73.

Henshaw, G.G., O'Hara, J.A. and stamp, J.A. 1985. Cryopreservation of potato meristems. In K.K. Kartha (ed), *Cryopreservation of Plant Cells and Organs.* CRC Press, Boca Raton. pp. 159-71.

Kartha, K.K. 1985. Meristem culture and germplasm preservation. In K.K. Kartha (ed), *Cryopreservation of Plant Cells and Organs.* CRC Press, Boca Raton. pp. 115-35.

Mix, G. 1985. Preservation of old potato varieties. In A. Shafer-Menuhr (ed), *In Vitro Techniques–Propagation and Long-term Storage.* Martinus Nijhoff/Junk Publishers, Dordrecht, Boston, Lancaster. pp. 149-53.

Towill, L.A. and Briggs, E.E. 1989. Techniques for preservation of plant germplasm. In L. Knutson and A.K. Stoner (eds), *Biotic Diversity and Germplasm Preservation–Global Imperatives.* Kluwer Academic Publishers, Dordrecht, Boston, London. pp. 379-405.

Withers, L.A. 1985. In K.K. Kartha (ed), *Cryopreservation of Plant Cells and Organs.* CRC Press, Boca Raton. pp. 243-69.

8

ROLE OF MAMMALIAN CELL CULTURES FOR THE PRODUCTION OF BIOLOGICALS

B. GRIFFITHS
PHLS-CAMR, Porton Down, Salisbury, England

INTRODUCTION

From the discovery by Enders (1949) that viruses could be grown in cultured cells *in vitro*, it took only a short time for the production of the first virus vaccine from cells in culture–the Salk Polio Vaccine, 1954. For the next 20 years, viral vaccines were the only cell-derived product.

Initially, vaccines were produced in primary monkey kidney cells, but their use was severely limited by both supply of material and safety considerations concerning their origin. Production of a biological for human and veterinary use automatically imposes compliance with safety regulations formulated by the relevant regulatory agency (e.g., FDA).

The derivation and acceptance of the human diploid cell line (HDC), WI-38, which was a safer and more reproducible and controllable substrate than primary cells, enabled a much wider range of vaccines to be developed from cell culture (Table 8.1).

TABLE: 8.1
LANDMARKS IN ANIMAL CELL BIOTECHNOLOGY.
***NB* VETERINARY VACCINES, E.G., FMDV, 2000 MILLION DOES PA SINCE 1970s.**

1954	polio vaccine
1960s	HDC vaccines–measles (1963), rabies (1964), mumps and rubella (1969)
1970s	new limited application vaccines, e.g., varicella & CMV (1974), TBE (1975)
1975	monoclonal antibody technique
1979	first recombinant cell line
1980	interferon, scale-up to 80001
1981	first MAB diagnostic kit
1982	first recombinant product (insulin)
1986	interferon licensed, recombinant cell products in trial
1987	OKT3 MAB licensed
1988	tPA licensed
1990	EPO licensed

Although this cell line was very acceptable from the licensing point of view, it was inhibiting further developments because manufacturers were unable to use truly large scale unit processes for a cell line that was anchorage-dependent and of low growth potential.

Animal Cell Culture has become a dominant technology in the past few years owing to a change in attitudes as well as developments in techniques. Whilst manufacturers were locked into the use of human diploid or primary cells by current licensing regulations, animal cell culture was never going to be a first choice production method. Manufacturers of human biologicals had always looked with envy at their veterinary counterparts using heteroploid (e.g., BHK) cells in 8000 1 stirred tanks to produce FMDV vaccine, and hoped that one day they might use a similar methodology. This moment has now arrived as three major biologicals manufactured from heteroploid, tumour-derived or recombinant, cell lines have been licensed in the past few years (Table 8.2). The reasons for products from such cell lines being accepted include major improvements in purification technology, and increased sensitivity of assay techniques for dangerous contaminants (e.g., DNA, retroviruses). Thus the process and final product can be properly validated. The acceptance of more realistic standards (e.g., for DNA) has also helped greatly.

This breakthrough which has broken the dogma of only using primary and HD cells has completely opened up the horizons for future developments in animal cell biotechnology. However important, it has not been the sole development over the past few years (Table 8.3). Scale-up to 8000 1 bioreactors for interferon, 2000

1 for monoclonal antiboy, and 4000 1 for microcarrier culture, has proved that cell culture can be operated at the industrial plant level. A further example of scale-up capability is that the total worldwide production of MAB in 1980 was 1 g, in 1990 this was 100 kg or more. Other developments that have proceeded successfully are the ability to grow cells to high density (50-fold increase in unit cell density), and in serum-free, low-protein media. The last two factors have considerably reduced the advantage gap that bacterial systems had over animal cells in terms of productivity and culture costs.

TABLE: 8.2
RECENTLY LICENSED PRODUCTS FROM HETEROPLOID CELL LINES. * LIMITED LICENCE (FRANCE, BELGIUM).

Product	*Cell line*	*Type*
Interferon	Namalva	Lymphoblastoid
OKT3 MAB	Hybridoma	Myeloma
tPA	rCHO	Recombinant
Polio vaccine	Vero	Heteroploid*
Rabies vaccine	Vero	Heteroploid*
Erythropoiten	rCHO	Recombinant

TABLE: 8.3
RECENT ADVANCES IN ANIMAL CELL TECHNOLOGY.

1.	Scale of production	–	IFN (8000 1)
		–	MAB (2000 1)
2.	Quantity of product	–	MAB 1 g (1980)
			100 kg (1990)
3.	Licensing of heteroploid cell products		
4.	Capability for high-density long-term culture		
5.	Development of serum-free (low protein) media		

CURRENT STATUS OF ANIMAL CELL BIOTECHNOLOGY

Animal cell bioreactors have always been a subject of topical discussion because of their shortcomings for industrial/commercial applications, or because the wide range that has developed (Griffiths, 1988) makes a choice difficult. The diversity in available bioreactors was inevitable given that both anchorage-dependent and suspension cells, growth-related and growth-independent products, and intracellular and extracellular products, have to be allowed for.

Also there are two strategies for scale-up: one is to increase the culture size but not its unit performance, the other is to increase unit performance but at a low volume. The challenge is to develop a universal system that has both volumetric and process intensity scale-up. How far we have advanced towards this goal is summarized below.

Scale-up–Overcoming Limiting Factors

The limiting factors in scale-up, and some of the solutions to overcome them, are summarized in Table-8.4. Initially scale-up was limited by oxygen, by being able to mix homogeneously fragile cells, and in the case of anchorage-dependent cells, to provide a large surface area for cell attachment.

Various solutions to these problems were found, including airlift instead of stirred mixing and for oxygenation, oxygenation by bubble-free membrane (Lehmann *et al.*, 1988) systems, and hollow fibre technology (Hopkinson, 1985; Tharakan and Chau, 1988). Having achieved some scale-up, the next limiting factor was nutrient limitation and build-up of toxic metabolites (ammonia being the most notorious).

TABLE: 8.4
A SEQUENTIAL SUMMARY OF FACTORS LIMITING CULTURE SCALE-UP, AND SOME OF THE SOLUTIONS THAT HAVE ENABLED FURTHER SCALE-UP IN VOLUME (V) OR CELL DENSITY (D) TO BE MADE. S-SUSPENSION CELLS; A-ATTACHED CELLS.

Limiting factor	*Solution*	*Action*
Oxygen (s)	membranes (V) hollow fibres (D)	bubble-free aeration oxygenation loops
Mixing (S)	airlift (V)	low shear
Surface area (A)	plates (V) ceramics (D) microcarriers (V)	
Nutrients/toxic metabolites (A, S)	perfusion (V)	hollow fibre loops (S) spin filters (A)
Cell retention (S)	immobilization (D) (immurement)	hollow fibres encapsulation membranes
Culture longevity V & D scale-up (A, S)	immobilization (V, D) entrapment	fluidized/fixed beds

This was solved by perfusion techniques, either *in situ* such as the spin filter (Griffiths *et al.*, 1987) or external hollow fibre loops (Emery *et al.*, 1987). Perfusion techniques have made a high impact on cell technology as they are the means of achieving high cell densities (suspension cells), or using more space-efficient substrates for cell attachment (to replace the headspace which constitutes 90% of the volume of a roller bottle for instance).

With this capability, a wide range of bioreactors based on hollow fibres (Tyo *et al.*, 1988; Schonherr *et al.*, 1987), membranes (Klement *et al.*, 1987) and ceramics (Berg and Bodeker, 1988) became available. To increase both productivity and scalability further, the techniques of cell immobilization has become a popular method.

IMMOBILIZATION

The advantages of immobilization are that it stabilizes cell in culture and enables longterm culture periods (months) to be achieved, and increases cell productivity by facilitating perfusion of suspension cells and increasing the usable substrate surface area for attached cells. Suspension cells can be immobilized by immurement or entrapment. Examples of immurement are hollow fibres (Hopkinson, 1985; Tharakan and Chau, 1988), membrane reactors (Scheirer, 1988) and encapsulation (Duff, 1985; Nilsson *et al.*, 1986) techniques. They do allow very high unit cell densities to be maintained (over 10^8 ml) but are a problem to scale-up beyond the 1-2 litre scale. An additional bonus is that by choosing the molecular weight cut-off of the membranes/fibres, one can compartmentalize the product in a concentrated form separate from the main bulk of the medium. Entrapment methods include porous ceramic cartridges (Opticell), fibres (Larsson and Litwin, 1987), sponges (Murdin *et al.*, 1987; Lazar *et al.*, 1988) and porous particles (Nilsson *et al.*, 1986). Although these methods do not permit such a high unit density (approximately 5×10^7/ml), they have greater scale-up potential and need less sophisticated equipment and control (Griffiths, 1990).

Anchorage-dependent cells are of course by definition immobilized. Thus the technology has been purely aimed at giving as high a substrate area, and thus cell density, as possible per unit volume. Although plate systems (Griffiths *et al.*, 1987) and packed beds of glass spheres (Whiteside and Spier, 1981) have been scaled up to several hundred litres, the most successful method for this type of cell is the microcarrier. Recently, the microcarrier has been developed for high-volume and high-density culture.

CELL CARRIERS (MICROCARRIERS)

The concept of cells growing as a monolayer on small particles which could then be treated in the same manner as a suspension cell originated with van Wezel (1967). This brilliant solution to achieve a truly unit scale-up method for animal cells, utilizing large-scale fermenter technology, had to wait the manufacture of suitable carriers before it became an acceptable alternative to multiple unit processes such as the roller bottle. The design of a microcarrier with the right surface and physico-chemical properties was initially achieved with the Cytodex (Pharmacia) range. Today there is a wide range of excellent microcarriers commercially available (Griffiths, 1986) which allows great flexibility in selecting for a particular cell type and for the required process parameters. Typically, microcarriers are used at concentrations which support up to 2×10^6 cells/ml, but spin filters are far more effective with 200 micron microcarriers than with 20 micron suspension cells and allow increases in process intensity up to about 3×10^7ml (Griffiths *et al.*, 1987; Looby and Griffiths, 1987).

A disadvantage of spherical microcarriers is that the surface-area:volume ratio is so low. However, the surface area has been enormously increased by using porous beads with fully interlinking passageways. This method was pioneered by Verax (Rundstadler and Cernek, 1988), followed by the Percell (Nilssson, 1988)and Siran systems (Looby and Griffiths; 1988, 1989).

With the correct pore diameter (approximately 2-3 times greater than the cell diameter), good cell and nutrient penetration is possible. The added advantage of a porous sphere is that it is equally advantageous for suspension (entrapment) and attached cells.

This in many ways is a distinct advantage over the immurement of suspension cells in a gel sphere (e.g., Damon system) in that a better exchange of nutrients and waste products occurs resulting in much higher cell viability for longer periods.

At the beginning of this paper, it was stated that the challenge has been to design a universal system with volumetric and process intensity scalability. How far does the porous microsphere achieve this objective?

In Table 8.5, its advantages are listed. Analogous systems are the stirred microcarrier system which is currently used at 1200 1 and 4000 1 scale, and the fixed bed of glass spheres at 200 1. Thus the potential for volumetric scale-up is there.

Cell yields of 2×10⁸ ml or better are being achieved (Looby and Griffiths, 1988), often with non-optimized conditions, so the ability to increase unit process intensity by 50-fold is proved (Table 8.6).

TABLE: 8.5

POROUS CARRIERS–ADVANTAGES COMPARED TO SOLID MICROCARRIERS AND LOW CELL-DENSITY SYSTEMS.

- Unit cell density 50-100-fold higher than free suspension
- Suitable for attached and suspension cells
- *In situ* 100-250-fold seed expansion eliminates scale-up steps
- Diffusion into a sphere efficient (30% penetration = 70% of volume)
- Suitable for stirred, fluidized and fixed bed bioreactors
- Good scale-up potential by comparison with analogous systems (e.g., microcarrier at 4000 1)
- Cells protected from shear
- Capable of long-term continuous culture
- 3-dimensional structure easily and beneficially derivitized

TABLE: 8.6

TYPES OF POROUS MICROCARRIERS, MODES OF USE, AND COMPARATIVE CELL YIELDS (X MILLION PER ML SETTLED BED VOLUME). A - DATA FROM AUTHOR'S LABORATORY, B - NOT ASSESSED BY AUTHOR, NA- CARRIER NOT OPTIMIZED FOR CULTURE MODE, X - DESIGNATED CULTURE.

Microcarrier	*Material*	*Cell line*	*Fixed bed*	*Fluidized bed*	*Stirred*
A Microsphere (Verax Corp)	Collagen	CHO	NA	120	NA
		Hybridoma	NA	198	NA
CULTISPHER G (HiClone)	Gelatin	CHO	NA	37	20
SIRAN (Schott Glaswerke)	Glass	CHO	58	78	NA
		Hybridoma	70	58	NA
B INFORMATRIX (Biomat Corp)	Collagen-glycosaminoglycan		NA	X	NA
MICROCARRIERS (Asahi Chemicals) (Kirin Cellsnow)	Cellulose		NA	?	X
POLYHIPE (Microporous Materials)	Polystyrene		NA	?	X

The fact that it is equally suited to suspension and attached cells has already been discussed, but the concept also can be utilized in stirred and fluidized, as well as fixed bed, reactors (Looby and Griffiths, 1988). It also fulfils the need for culture longevity and stability.

A limiting factor in any matrix is the efficiency of nutrient exchange, i.e., diffusion path lengths, etc. The sphere is ideal in that 70% of its volume occupies the outer 1/3rd diameter, i.e., diffusion paths can be vary short.

In conclusion, it is too early to state that this is the *Universal System* we have all been waiting for, but it is certainly a highly probable, and currently the only candidate for this title. It is interesting to note that immobilization of micro-organisms is an old technique, but because of biomass blockage and low oxygen transfer it has never become a successful method. Because cells secrete their product and have a lower oxygen requirement the use of immobilization is now far more advanced for animal than bacterial cells.

BIOREACTOR PRODUCTIVITY

A number of bioreactors have been described, all of which have advantages and disadvantages. The question is how do they perform in comparative terms. The data in Table 8.7 show clearly that a perfused porous microcarrier system, e.g., Verax, supports a significantly higher cell density, specific productivity, and bioreactor productivity than other culture systems. Performance can be measured in many ways but basically we are interested in the unit number of cells it will support, and the product secretion rate of those cells. It is always helpful to have a target, but what do we choose as a realistic batch cell number? A suggestion is to use the human body as a model where over 10^{13} cells are physiologically supported in a compact, well regulated, and efficient manner. In Table 8.8 various examples of bioreactor are listed with typical cell number they can support. Conventional technologies of stirred (up to 8000 1) and airlift (2000 1) will, if big enough (i.e., over 5000 1), achieve this target, but they are mainly batch systems which need a huge investment in plant and back-up facilities. Productivity can be increased using perfusion with a spin filter, but this has only been proven at 100 1 to date (Griffiths, 1988).

Encapsulation is difficult to scale-up and the high cell densities quoted are for the intracapsular volume which is only 20% of the total volume. High cell density systems (fibres, membranes, etc.) cannot be volumetrically scaled up, but do allow continuous processes. The porous bead, so far operated at the 25 1 scale (Rundstadler and Cernek, 1988), does come near to the target cell density, and does have all the advantages listed in Table 8.5, plus less plant investment than the high-volume systems. One conclusion from Table 8.8 is that if 10^{13} cells is accepted as a sensible target, then the culture systems are available to achieve it.

TABLE: 8.7
COMPARISONS OF MAB PRODUCTIVITY IN VARIOUS CULTURE SYSTEMS.

System	*Active volume*	*Total cell no. ×10^7*	*Max. cell no. 10^{10}/l*	*Product. avge cell mg/l/d*	*Max. specific product μ/10^6/d*	*Product. per month mg/l*
T-flask	100	14.7	0.15	9.3	18.0	140
Verax syst 1	10	300.0	15.00	540	32	16200
E-flask	10	35.0	2.08	160	22	2400
Stirred	1000	140.0	0.19	34	16.5	510
Chemostat (.6day)	1000	120.0	0.12	24.5	20.4	736
Airlift	5000	600.0	0.12	18.5	17.0	277
Fixed bed	1000	4000.0	2.75	166	22.0	4980

TABLE: 8.8
COMPARATIVE CELL PRODUCTIVITIES IN VARIOUS BIOREACTOR SYSTEMS. * REALISTIC SCALE-UP. NB MICROCARRIER AT 1200 1.

HI Vol/Lo Density			
Stirred tanks	10000	3	300
Airlift	2000	3	60
Lo Vol/HI Density			
Encapsulation	50	20	10
Hollow fibre	1	200	2
Membroferm	3	200	6
HI Vol Modifications			
Spin filter	100	50	50
Centrifugation	200	30	60
Porous Microcarriers			
Current scale	25	80	20
Probable scale*	250	80	200

CELL PRODUCTIVITY

Having set a target for cell numbers, we need to do the same for expression of a product from cells. If we take 250 mg protein per million cells as an average, then what percentage of this protein should we expect to be secreted as a specific product? Cells can double in 24 hr or less, then they have the biosynthetic capability of producing 250mg protein per day. Current production rates are, for a batch culture, on average over 50 mg/10^6 cells/4-6 days, i.e., about 10 mg/10^6 cells/day. This is only 4% of the cell protein, and is obviously a production rate which must be increased. It is possible to achieve higher productivities per day in some systems (see review Merton, 1988), even up to 1500 mg/10^6 cells. However, such high rates cannot be maintained for very long in culture (Murakami, 1990). So what is a realistic target production rate? Murakami suggests that 30% (of the cell protein) is achievable by cultured myeloma cells, and compares this with mature B cells in lymph nodes which secrete nearly 50 mg 1gG/10^6 cells/day.

In reviewing bioreactors, the stated aim was for a system supporting 10^{13} cells. With a conservative product expression rate of 50 mg/10^6 cells/day such a culture would yield 500 g of product per day. This would appear to be both achievable and an economically acceptable production rate. However, the productivity of MPC-11 myeloma cells (720g Igb/10^{11} cells/day; Murakami and Yamada, 1987) shows the potential is there to increase product expression levels further. A recent example of recombinant CHO cells expressing up to 200 mg/1 of hGH (Griffiths, 1990) is proof that expression levels will be improved. What is now needed is the technology to maintain cells at this high productivity rate over long culture periods. This will be achieved by attention to the physiological and biochemical balance of the growth medium and by providing the correct microenvironment.

CONCLUSIONS

Animal cell culture is a more exacting and expensive production method than bacteria, so can never be a first choice method. However, many of the complex molecules we need to produce cannot be manufactured in an active form by bacteria. Post-translational changes, e.g., glycosylation, cannot be carried out by bacteria, nor in many cases can the correct 3-dimensional configuration be correctly expressed. Thus animal cell culture becomes the method of necessity, if not of choice, for many complex molecules. Other developing technologies which may partially replace the need for using animal cells are synthetic and transgenic products, and it is likely that each technology will find its own niche. The use of porous microcarriers has given manufacturers the option of having a high cell density system that is scalable. The advantages are listed in Table 8.5, and the question of being able to license a product from a continuous process has been addressed and guidelines prepared (US FDA, 1987). The advantages of a high-density perfusion process (Table 8.9) make porous microcarrier technology an ideal scale-up process.

If animal cell culture technology is to remain a dominant technology, then further improvements are needed. This will be done by a combination of genetic modification, improved knowledge of cell physiology, and in bioreactor design. These factors will be heavily interlinked, for instance the use of specific growth factors will control gene expression; the provision of an effective micro-environment (for instance 3-dimensional microcarriers derivitized with extracellular matrix protein) will allow culture longevity and maximize product expression.

TABLE: 8.9
ADVANTAGES OF HIGH CELL DENSITY PERFUSION CULTURE

- Smaller reactor volume (NB medium tanks)
- More concentrated product–higher specific productivity
- Reduced serum/growth factor requirement
- Reduced occurrence of nutrient/toxic inhibition
- Smaller facilities–quicker start-up
- *In situ* seed expansion (100-250-fold)
- Long-term (continuous) process
- Reduced product exposure and cell debris accumulation

REFERENCES

Berg, G.J. and Bodeker, B.G.D. 1988. Employing a ceramic matrix for the immobilization of mammalian cells in culture. In R.E. Spier and J.B. Griffiths (eds), *Animal Cell Biotechnology.* Academic Press, London. Vol. 3, pp. 321-25.

Duff, R.G. 1985. Microencapsulation technology: a novel method for monoclonal antibody production. *Trends Biotechnol.*, 3, 167-70.

Emery, A.N., Lavary, M., Williams, B. and Hander, A. 1987. Large-scale hybridoma large culture. In C. Webb and F. Mavituna (eds), *Plant and Animal Cells; Process Possibilities.* Ellis Hallward, Chichester, England. pp. 137-46.

Enders, J.F., Weller, T.H. and Robbins, F.C. 1949. *Science,* 109, 85.

Griffiths, J.B. 1986. In R.I. Freshney (ed), *Animal Cell Culture: A Practical Approach.* IRL Press, Oxford, England. pp. 33-70.

Griffiths. J.B.1988. In R.E. Spier and J.B. Griffiths (eds), *Animal Cell Biotechnology.* Academic Press, London. Vol. 3, pp. 179-220.

Griffiths, J.B. 1990. In R.E. Spier and J.B. Griffiths (eds), *Animal Cell Biotechnology,* Academic Press, London. Vol.4, pp. 150-60.

Griffiths, J.B., Cameron, D.R. and Looby, D. 1987. A comparison of unit process systems for anchorage-dependent cells. *Dev. Biol. Standard,* 66, 331-38.

Hopkinson, J. 1985. Hollow fibre cell culture systems for economical cell-product manufacturing. *Bio/Technology*, 3, 225-30.

Klement, G., Scheirer, W. and Katinger, H.W.D. 1987. Construction of a large scale membrane reactor system with different compartments for cells, medium and product. *Dev. Biol. Standard*, 66, 221-26.

Larsson, B. and Litwin, J. 1987. The growth of polio virus deployed fibroblasts grown with cellulose microcarriers in suspension. *Dev. Biol. Standard*, 66, 385-90.

Lazar, A., Silberstein, L., Mizrahi, A. and Reuveny, S. 1988. An immobilized hybridoma culture perfusion system for production of monoclonal antibodies. *Cytotechnology*, 1, 333-38.

Lehmann, J., Vorlap, J. and Buntemeyer, H. 1988. In R.E. Spier and J.B. Griffiths (eds), *Animal Cell Biotechnology*. Academic Press, London. Vol 3 pp. 221-37

Looby D. and Griffiths, J.B. 1987. A comparison of oxygenation methods in a 40 1 stirred bioreactor.. In R.E. Spier and J.B. Griffiths (eds), *Modern Approaches to Animal Cell Technology*. Butterworth Scientific, Guildford, England. pp. 449-53.

Looby, D. and Griffiths, J.B. 1989. Fixed bed porous glass sphere (porosphere) bioreactors for animal cells. *Cytotechnology*, 1, 339-46.

Looby, D. and Griffiths, J.B. 1989. Immobilization of animal cells in fixed and fluidized porous glass sphere reactors. In R.E. Spier and J.B. Griffiths (eds), *Modern Approaches to Animal Cell Technology*. Butterworth Scientific, Guildford, England. pp. 336-44.

Looby, D. and Griffiths, J.B. 1990. *Trends Biotechnol.*, 8, 204.

Merton, O-W 1988. Batch production and growth kinetics of hybridomas. *Cytotechnology*, 1, 113-21.

Murakami, H. 1990. What should be focused in the study of cell culture technology for production of bioactive proteins. *Cytotechnology*, 3. In press.

Murakami, H. and Yamada, K. 1987. Production of cancer specific monoclonal bodies with human hybridomas and serum-free high-density, perfusion culture. In R.E. Spier and J.B. Griffiths (eds), *Modern Approaches to Animal Cell Technology*. Butterworth Scientific, Guildford, England. pp. 52 -76.

Murdin, A.D., Thorpe, J.S. and Spier R.E. 1987. Immobilization of hybridomas in packed-bed bioreactors. In R.E. Spier and J.B. Griffiths (eds), *Modern Approaches to Animal Cell Technology*. Butterworth Scientific, Guildford, England. pp. 420-36.

Nilsson, K, 1988. Macroporous gelatin microcarriers for animal cell culture. *BioMedia (Biolytica)*, 1, 13-15.

Nilsson, K., Buzasaky, F. and Mosbach, K. 1986. Growth of anchorage-dependent cells on macroporous microcarriers. *Bio/Technology*, 4, 989-90.

Runstadler, P.W. and Cernek, S.R. 1988. Large scale fluidized-bed immobilized cultivation of animal cells at high densities. In R.E. Spier and J.B. Griffiths (eds), *Animal Cell Biotechnology*. Academic Press, London. Vol 3, pp. 305-20.

Scheirer, W. 1988. high-density growth of animal cells within cell retention fermenters equipped with membranes. In R.E. Spier and J.B. Griffiths (eds), *Animal Cell Biotechnology*. Academic Press, London. Vol. 3, pp. 263-81.

Schonherr. O. T., van Gelder, P.T.J.A., van Hees, P.J., van Os, A.M.J.M. and Roeloffs, H.W.M. 1987. A hollow fibre dialysis system for *in vitro* production on monoclonal antibodies replacing *in vitro* production in mice. *Dev. Biol. Standard.* 66. 211-20.

Tharakan, J.P. and Chau, P.C. 1988. A radial flow hollow fibre reactor for the large scale culture of mammalian cells. *Biotechnol. Bioeng.*, 28, 329-42.

Tyo, M.A., Bulbulian, B.J., Zaspel, B. and Murphy, T.J. 1988. large-scale mammalian cell culture utilizing ACUSYST technology. In R.E. Spier and J.B. Griffiths (eds), *Animal Cell Biotechnology*. Academic Press, London. Vol. 3, pp. 357-71.

Van Wezel, A.L. 1967. Growth of cell strains and primary cells on microcarriers in homogeneous culture. *Nature (London)*, 216, 64-65.

Whiteside, J.P. and Spier, R.E. 1981. The scale-up from 0.1 to 100 litre of a unit process system based on 3 mm diameter glass spheres for the production of four strains of FMDV from BHK monolayer cells. *Biotechnol. Bioeng.*, 23, 561- 65.

US Dept. of Health and Human Services, FDA, Office of Biologics Research and Review. 1987. Points to consider in the characterization of cell lines to be used to produce biologicals.

9

THE ROLE OF PLANT CELL CULTURES IN BIOTECHNOLOGY

H.M. SCHUMACHER
DSM—German Collection of Micro-organisms and Cell Cultures, Braunschweig, Germany

Summary. The first experiments to culture isolated plant cells *in vitro* was made by Haberlandt in 1902. From these first attempts on many different culture. Systems, a wise range of applications for these *in vitro* techniques were evaluated.

In 1958, Reinert in Germany and Stewart in the USA succeeded in regenerating a whole intact plant from an undifferentiated callus culture.

This event opened up the field of agriculture for plant cell culture technology, and today it is almost impossible to distinguish clearly between agricultural and biotechnological applications.

Today, certain *in vitro* culture systems support conventional techniques and provide new breeding techniques for the improvement of crop plants.

The *in vitro* culture of excised immature embryos to maturity allows natural barriers for crossings to be overcome. The normal degree of variation of a callus culture can–after regeneration of plants–be used to enhance the naturally occurring variation for crossings.

Somatic hybridization of protoplasts isolated from cultured cells or plants opens up the way for crossings between different species or even plants from different genera. Pollen culture and subsequent regeneration of plants cell cultures provide a method for achieving complete homozygosity within one generation.

At present, the most important cell culture application concerning economical aspects is the mass propagation of elite plants by meristem or shoot-tip culture.

Apart from agricultural applications, efforts have been made to produce plant raw material by *in vitro* culture of plant cells. These efforts focused mainly on the production of pharmaceutical plant compounds by suspension cultures. Plants still provide a good number of medical compounds. Many of them like the anti-cancer drugs vinblastine or taxol are structurally too complex to be produced by chemical synthesis.

In Japan, the anthraquinone shikonin has been produced by plant cell cultures and in Germany the production of immunostimulant polysaccharides is being planned.

Other targets for the production by cell culture techniques are plant enzymes or flavours and dyes for the food industry.

Nevertheless, in many cases it transpires that a desired compound is not produced by undifferentiated plant cells.

A novel approach to overcome this problem is the *in vitro* culture of differentiated plant organs. A certain hormone regime allows the formation and culture of roots or shoots *in vitro.*

INTRODUCTION

The importance of plants and plant products for human culture has been primary since the beginning of civilization. They had and still have uses that extend beyond food production.

In former times, nearly all raw materials for technical processes have been taken from plants. Houses and vehicles were made from timber, clothes from cotton, and oils, dyes, and resins extracted from plants provided the starter material for early chemical processes. Until 1890, the health care system also had to rely totally on plant-derived drugs. The main reason for the broad technical application of plants is their versatile and fascinating secondary metabolism. No other group of organisms is capable of creating such an enormous number of secondary metabolites which in many cases have a high biological activity.

Even today in Germany, 12% of the chemically-defined substances used for chemical treatment are plant-derived. Besides simple and cheap substances like caffeine, there are high valuable compounds like digoxin or vinblastine. Other fields of application for plant secondary metabolities are the cosmetics and food industry. Market surveys show that food technology and pharmacy covers worldwide three-quarters of all biotechnology sales. This illustrates the importance plant products may have for this field. The trend that customers prefer natural colours, flavours, and medicines will probably increase this importance. Plant cell cultures may provide new tools for the biotechnological production of plant products. But they can also serve as tools for improving existing plant cultivars or establishing new cultivars for the production of new products. The consequence is that we cannot draw a distinct borderline between plant biotechnology and conventional agriculture. Both aspects of plant-tissue and cell-culture technology should therefore be mentioned.

HISTORY OF PLANT TISSUE AND CELL CULTURE TECHNOLOGY

A short summary of the history of *in vitro* culture techniques will illustrate the different types of tissue and cell cultures. First experiments to culture isolated plant cells *in vitro* were carried out by Georg Haberlandt in 1902. He had mainly scientific interests and wanted to prove the totipotency of single cells and thereby verify the theory of Schleiden and Schwann. Although he simply used Knop's salt solution, he was able to achieve cell growth and survival for several weeks. Unfortunately he failed in inducing cell divisions.

The first successes were not made with isolated cells but with plant organs. Only a few years after Haberlandt's first attempts, Hanning isolated embryos from the seeds of *Cruciferae* and cultured them under aseptic conditions also in a simple medium consisting of Knop's salt solution and sucrose. He was able to germinate the embryos and to grow plants from the seedlings. Also very old are efforts to grow isolated roots. First experiments were performed by a student of Haberlandt in 1924, but the first real root culture were achieved by White. He was the first to add B-vitamins to a cell culture media. Further progress of *in vitro* techniques was the first successful culture of isolated shoot-tips by Ball in 1946. Although progress was made with the culture of plant organs, for a long time all efforts to achieve a real culture of isolated cells remained unsuccessful. It was not until 1939 that Gautheret and Nobecourt managed, independently, to establish the first continuously-growing callus culture consisting of undifferentiated plant cells. They were able to induce cell division by adding auxin to their media. Auxin had been discovered shortly before and was the first known plant hormone. When it became possible to grow plant cells submerged, like micro-organisms, in 1954 the first suggestions were made to use these cultures for the production of secondary plant metabolites.

It was 54 years after Haberlandt's first *in vitro* culture experiments, that Reinert in Germany and Stewart in the USA reached his initial goal. By changes in hormone composition of the medium, they achieved the regeneration of an intact plant from somatic cells of a *Daucus carota* cell culture. This event was not only the proof for the totipotency of single plant cells but it opened up the route from cultured cells back to the plant.

Another important technology is the culture of isolated gametophytes, especially pollen. In 1966, for the first time, haploid plants were regenerated from cultured anthers (see Bhojwani and Razdan, 1983).

In the early 70s, technologies were developed to culture isolated dedifferentiated plant cells like calli or submerged and nearly all kinds of plant organs *in vitro*. The possibility of manipulating different plant organs and isolated plant cells and to regenerate intact plants from these cells and tissues led to a number of applications. We can distinguish four main fields:

1. the improvement of plant breeding techniques to achieve new plant varieties or products
2. the mass propagation of elite plants
3. the *in vitro* mass production of biotransformation of plant products by cultured plant cells
4. the use of plant cell cultures for screening procedures in applied research.

APPLICATIONS FOR PLANT BREEDING

Different *in vitro* techniques today can support conventional breeding methods and in some cases even provide quite new methods for plant breeding.

Embryo Culture

The oldest *in vitro* techniques which has been used for practical application is embryo culture. First experiments were carried out in 1904 and already by 1925 Laibach was using the method to cross different *Linum* species which normally do not form germinable seeds.

He succeeded in isolating immature embryos, grew them to maturity, achieved germination, and was able to grow plants.

These plants cannot be obtained by conventional crossing. Improved methods currently allow even the culture of immature tomato and barley embryos (Bhojwani and Razdan, 1983).

Production of Homozygous Plants

To generate a new plant variety that is stable during propagation via seed normally needs many backcrossings. The use of plants with a high degree of homozygosity reduces the necessary number of backcrosses. Therefore, a very important goal for plant breeding is to enable the generation of plants as homozygous as possible. *In vitro* techniques offer a way to achieve complete homozygosity within one generation. Haploid plants can be produced by the culture of pollen. The normal degree of ploidity can be restored by treatment with colchizin.

But pollen-culture-generated haploid plants themselves can be useful tools for plant breeding: otherwise recessive mutations that cannot be seen phenotypically can be sorted out directly with haploid plants. With *Datura* plants, haploid-level selection procedures have been successfully applied to enhance the plant scopolamin content (Bollon and Raquin, 1987).

Somatic Hybridization

The term somatic hybridization means the fusion of protoplasts derived from somatic cells. Fusion of isolated protoplasts can be achieved by two major techniques: (i) cell fusion induced by treatment with chemicals, (ii) electrofusion. The most frequently chemical means are treatment with polyethylene glycol (PEG) and treatment with high pH values and Ca^{2+} ions. The result is a hybrid cell with mixed cytoplasm, mixed plastids and mitochondria and–after fusion of the nuclei–mixed nuclear genome. Since protoplasts can be cultivated and regenerated to intact plants, this method provides a new *in vitro* technique for crossings between plants of different species and maybe even different genera. These 'crossings' would be impossible by sexual means. It also allows 'crossings' between different cultivars of vegetatively propagated plants. By microinjection it is also possible to mix the nuclear and the extranuclear genome (mitochondria, plastids) of different plants.

A critical point is the regeneration of intact plants from fusion products. Techniques for the regeneration of hybrid cells are not always available, and sometimes cells seem to lose their embryogenic potential.

Currently, successful practical application of these methods is to a large extend restricted to the *Solanaceae*. In the field of food-producing plants, potatoes in particular have been targets for plant improvement but even the secondary metabolism could be influenced in the case of *Datura* and *Atropa* plants (Czygan, 1984; Dixon 1987).

Somaclonal Variation

The basis for plant improvement via breeding is the naturally occurring innerspecific diversity. Diversity is also a phenomenon occurring among the cells of a cell culture. Very often, different cells exhibit different chromosome numbers of different metabolic potential.

In many cases it is unknown whether this is only a phenotypic diversity or if it has a genetic basis. Nevertheless, plants regenerated from calli show a certain degree of genetic variation termed somaclonal variation. The phenomenon of somaclonal variation can be used to enhance the naturally occurring innerspecific diversity for breeding programmes.

Furthermore, the possibility of exposing a high number of cultured plant cells to a certain selection pressure offers a way of selecting out cells with desired traits and to regenerate plants with the same traits.

By adding certain herbicides to the cell culture medium, it was possible to select out herbicide resistant cells and to regenerate herbicide resistant plants. The same method has also been applied to achieve plants resistant to fungal toxins or heavy metals as well as salt, cold- or drought-resistant plants with major crops like potato, sugarcane, tobacco (Evans and Sharp, 1986).

IN VITRO METHODS FOR THE MASS PROPAGATION OF PLANTS

Besides the described methods for improvement of plant cultivars, *in vitro* culture techniques can also be used for the mass propagation of plants. Two principally different techniques are used.

Today, the most frequently used technique makes use of the shoot-tip culture technique developed by Ball. Excised shoot tips can be cultured and grown to small plantlets. Adding cytokinins to the growth medium leads to branching by growth of adventitious buds. The branches can be cut off and grown to new plantlets. The same procedure achieves a further multiplication. Plantlets can be rooted by exposure to auxin-containing media and after an adaptation phase in the greenhouse, planted out in the field.

The other method makes use of the embryogenic potential of cell cultures. Cells are first propagated on the culture level in fermenters. Changes in the medium content, especially reduction of auxin content, can initiate embryo development. Embryos can be planted out on filter paper or agar, and grown to plants.

Although conventional propagation techniques via seeds or tubers are much cheaper, cell culture methods have certain advantages. They were first applied with high valuable ornamental plants like orchids.

A major advantage is that the methods allow the production of a virus free plant stock for plantations. This is especially important for long-term plantations, for example, of fruit trees.

Another advantage is that *in vitro* propagation allows the fast multiplication of elite plants vegetatively. This is especially important in combination with the methods for plant improvement mentioned above, and also for new transgenic plants.

The traits of elite plants achieved by these methods in many cases are not genetically stable when the plants are propagated via seeds. Nevertheless, *in vitro* propagation allows the fast mass production of such plants.

Although it implies the danger of producing a heterogeneous plant pool (by somaclonal variation), the propagation method *via* somatic embryogenesis might offer another elegant new technique for plant propagation. Somatic embryos have successfully been immobilized in certain matrices and it is suggested that this method be used for the production of artificial seeds.

All the methods mentioned so far are normally considered to be agricultural and not biotechnological methods. Nevertheless, we should note that they are not only applied to increase food yield, but also to improve the production of compounds for technological purposes.

A very interesting aspect for biotechnology is that these techniques will shorten the time that is necessary to improve and cultivate a wild-type plant that has never been used before (Bhojwani and Razdan, 1983).

SECONDARY METABOLITE PRODUCTION BY PLANT CELL CULTURES

Direct Production with Plant Cell Cultures

Soon after the first suspension culture was established, suggestions were made to produce plant compounds by cell cultures. A lot of different secondary plant metabolities like lignans, flavanoids, quinones, and terpenes have been found in plant cell cultures. More than 100 different tryptophane-derived alkaloids could be detected. Many of them have interesting biological activities. The most important need to make a cell culture process profitable is to increase the product yield. Many different techniques have been developed for this purpose. In some cases, product formation could be increased by changes of the medium composition. The yield of the alkaloid raucaffricine in cell cultures of *Rauwolfia serpentina* depends on the carbon source and its concentration. A further approach is the use of a specific production medium different from the growth medium. With cell cultures of *Lithospermum erythrorhizon*, the product yield increased 18-fold by optimizing the growth conditions (Fontanel and Tabata, 1987). Also the feeding of specific precursors can increase the formation of a desired end product.

In some cases, the application of microbial elicitors increases the yield of specific antimicrobial plant compounds, so-called phytoalexines. The production of such metabolities is often initiated only in the presence of elicitors. In Canada, efforts are made to produce the alkaloid sanguinarine by this method (DiCosmo, 1990).

The most important way of increasing product yield is to establish high-yielding cell culture lines by selection procedures. This approach takes advantage of the already-mentioned heterogeneity of the cell population in a newly established cell culture. Different techniques for cloning of plant cells using feeder cells or conditioned media have been developed to carry out selection procedures. If the desired compound is coloured or shows a certain fluorescence, high-producing cells or cell clusters can be detected optically and picked up directly. For other substances, immunological detection methods were designed to screen high numbers of cloned cell colonies. Even application of commercial cell sorters is possible.

Since different cell culture strains tend to react differently even to changes of the medium composition or to the application of elicitors, cell selection might be the most important of the described methods. Unfortunately, in some cases selected high-yielding strains have turned out to be unstable when the selection pressure is not maintained (Fontanel and Tabata, 1987).

Although the contents of several secondary metabolites could be increased to a level far exceeding that of whole plants, other plant compounds are not formed at all by undifferentiated plant cells. These include commercially interesting substances like morphine, digoxin, or atropin. It turned out that the formation of these compounds is strictly bound to cell differentiation. Only if somatic embryogenesis is initiated do *Digitalis* cells start to produce cardiac glycosides. While undifferentiated cell cultures of *Cinchona* only produce anthraquinones in root cultures, the alkaloid quinine is formed again. Since it is still not possible to initiate product formation without inducing differentiation, in these cases efforts are made to produce those substances by organ or tissue culture.

In many cases, shoot formation can be achieved simply by exposing cell cultures to hormone-free media. Genetically transformed shoot-like structures can be obtained by infection with a mutant of *Agrobacterium tumefasciens*. It has been shown that such structures form the same terpene pattern as normal peppermint plants (Spencer *et al.*, 1990).

At present, the so-called 'hairy root' cultures are more important than shoot cultures. They can be obtained by infection of plant material with *Agrobacterium rhizogenes.* The microbe transforms the plant cells genetically in a similar way to *Agrobacterium tumefasciens.* It induces root formation at the infection side instead of tumour growth. After treatment with antibiotics, these 'hairy roots' can be cultured under aseptic conditions even in liquid media and without addition of plant hormones. The transformed root cultures show vigorous growth and exhibit a secondary metabolism that resembles that of normal roots. Pharmaceutically important compounds like hyoscyamine and atropin are formed in such cultures. In the Netherlands, it is planned to produce nematocidal thiophenes from *Tagetes* plants with 'hairy roots' (Adamse, 1990). Although roots are more difficult to grow than suspensions, large-scale fermenters up to 500 1 volume have already been developed (Wilson *et al.*, 1990).

Cell Cultures as a Part of Complex Processes

Cell cultures have also been used to carry out specific reactions in a multistep production process for certain compounds. One of the most interesting substances for large-scale production with plant cell cultures is the anticancer drug vinblastine. At the moment, the drug is isolated from *Catharanthus roseus* plants. Unfortunately, the dimeric indole alkaloid is not formed by cell cultures. However, cell cultures produce the monomers. In this case, a process has been designed where cell cultures are used for the production of at least one monomer and for the isolation of an enzyme, which is capable of dimerizing the monomers (DiCosmo, 1990).

The crucial step in the biosynthesis of the complex structure of the indole alkaloids is the formation of strictosidine from trytamin and secologanin. The catalyzing enzyme could be isolated from cell cultures and immobilized. A column with enzyme bound to a gel matrix could be used for an enzymatic strictosidine production. The genes for this enzyme could be successfully expressed in bacteria. These bacteria can carry out a biotransformation to yield strictosidine. In a screening project, chemically synthesized derivatives of the intermediate stritosidine have been tested for their pharmacological activity.

Biotransformations can also be carried out directly by plant cells. The most elaborate example for a biotransformation is the hydroxylation of β-methyldigitoxin to β methyldigoxin by *Digitalis lanata* cells. None of the substances is formed by undifferentiated plant cell cultures but when β-methyldigitoxin is added to the medium, they can hydroxylate it. Since digoxin is the more valuable product while digitoxin occurs as waste during digoxin isolation, this biotransformation can be used to make digoxin isolation economically more effective.

CELL CULTURES IN APPLIED RESEARCH

Apart from biotechnology, plant cell cultures make an extraordinary contribution to fundamental botanical research. For many investigations they are a much more comfortable experimental tool than intact plants. For some methods–for example, *in vitro*, NMR studies–they are even the only applicable tool. They played an important role for the elucidation of biosynthetic pathways and for the investigation of enzymatic reactions and the isolation of enzymes.

For applied research plant cell cultures can be interesting tools as well especially to carry out screening procedures. They have been used to finding new substances with biological activity. A number of substances like the paniculides or the dimeric indole alkaloid voafrin have been isolated from plant cell cultures for the first time. During recent years, more then 85 new substances, among them 23 new alkaloids, 19 terpenes and 39 quinones, have been isolated from not more than 30 cell cultures. These substances could not be detected in the whole plant (Ruyter and Stockigt, 1989). In general, cell cultures are easier to handle in a screening procedure than whole plants. In the case of rare plants, cell cultures can also provide the tool to carry out medium-scale production of a compound for further investigation.

A promising new anticancer drug is the terpenoid compound taxol, isolated from the bark of the yew tree, but this rare and slow-growing tree cannot provide enough material to carry out the clinical trials for this substance (Stafford and Fowler, 1991).

Cell cultures have also been used to study the biodegradation of xenobiotics. It has been shown that the biodegradation of compounds like herbicides, fungicides, or insecticides is carried out by plant cell cultures in the same way as by intact plants.

However, for the investigation of such degradations, plant cell cultures again are a better tool than plants. Here also they can serve as tools for medium-scale production of the degradation products for toxicological studies. Photoautotrophic cell cultures growing without carbon source in the medium will be of special importance for these purposes (Barz et al., 1990).

FUTURE APPLICATIONS

The broadest commercial application of *in vitro* culture techniques presently is the field of micropropagation. Today it is not only applied to high valuable ornamental plants but also to several crop plants. The major advantages are the fast multiplication of new varieties and the production of virus-free plants. In 1986, more than 130 commercial laboratories worked in this field, and by 1990 300-500 million *in vitro* propagated plants were produced in the Netherlands alone. It can be estimated that with the creation of new varieties by cell culture methods or genetic engineering, the importance of this technology will certainly increase. It offers a way to bring new plants for the production of new substances onto the market in a much shorter time than would be possible by conventional agricultural methods.

Although great efforts have been made in the past, up to now only few examples of large-scale production of classical secondary plant metabolite with cell, tissue, or organ culture have been realized. In most cases, the production by plant cultivation turned out to be cheaper than a competing cell-culture process. The high costs will restrict further efforts even in the future in the production of high valuable fine chemicals, food additives, dyes or pharmaceuticals. In nearly all cases where cell culture methods had a chance to be applied commercially, specific problems for the production of the targets with conventional plant cultivation occurred. In the classic case of shikonin, plant roots grow very slowly and contain only 1-2% of the substance. In the case of vinblastine, the plant content is even lower (0.01% dry weight) (Stafford and Fowler, 1991). With taxol, conventional methods are not even able to supply enough material for clinical trials. The world's biggest fermentation facility for plant cell cultures has been built in Hamburg. The DIVERSA corporation intends to produce immunostimulant polysaccharides from *Echinacea purpurea* cell cultures in this plant. The target polysaccharides for this process turned out to be the most active ones that could be isolated, but they could be isolated only from cell cultures in considerable yields and an alternative production by plant cultivation is not possible (Westphal, 1990). Cost estimations show that investments for large-scale production facilities make the most important contribution to the costs of the whole process. These investments can only be justified by the production of highly valuable compounds and it can be expected that classical secondary plant metabolites which can easily be produced by agricultural means will be produced by plant cultivation in the future.

On the other hand, cell culture technology will be the method of choice whenever a new technology for the production of high valuable compounds is needed. This is especially true for substances which have never been used before, but also for pharmaceuticals which are still extracted from plants collected in the wild. In many cases, cell-culture technology will be able to develop a new technology for the production of a new compound faster than conventional agriculture. New targets for cell culture production might be new secondary metabolites but also specific plant enzymes. Molecular biology already recognized plant cell cultures as a suitable system for the production of mammalian proteins (Stafford and Fowler, 1991). Plants are even capable of producing intact monoclonal antibodies, and cell cultures of these plants may offer a cheaper and safer way for their production than animal cell cultures.

Increased knowledge about biosynthetic pathways and the modern techniques of molecular biology may revolutionize even the field of secondary metabolism in the future. Zenk and coworkers showed that the formation of berberine is carried out in nine specific enzymatic reactions. When only one of these enzymes is missing, the end product is no longer formed (Galneder and Zenk, 1990). Genetic engineering might be able to repair such deficiencies (Yamada and Hashimoto, 1990). On the other hand,'anti-sense' mRNA technology might be able to set certain blocks in a biosynthetic pathway. This might result in the accumulation of intermediates of even new products that have never been tested for their biological activity in the past.

During the process of industrialization, the importance of plants as raw material for technical processes dwindled, to a large extent replaced by mineral resources like metals, coal, mineral oil and its chemical derivatives. Nevertheless, the importance of plants for food production and as a source of drugs and fine chemicals remained.

Increased knowledge about the ecological importance of secondary metabolites or enzymes and their highly specific pharmacological, antimicrobial, or herbicidal activities will increasingly result in the replacement of relatively unspecific synthetic compounds by more specific and more effective compounds of biological origin. For compounds of plant origin, plant cell cultures will be the preferred method for their production in many cases.

REFERENCES

Adamse, P. 1990. Selection of high-yielding cell lines of *Tagetes* using flow cytometry. In H.J.J. Nijkamp, L.H.W. van der Plas and J. van Aartrijk (eds), *Progress in Plant Cellular and Molecular Biology.* Kluwer Academic Publishers, Dordrecht, The Netherlands.pp. 726-32.

Barz, W., Jordan, M. and Metschulat, G. 1990. Bioconversion of exenobiotics (pesticides) in plant cell cultures. In H.J.J. Nijkamp, L.H. W. van der Plas and J. van Aartrijk (eds), *Progress in Plant Cellular and Molecular Biology.* Kluwer Academic Publishers, Dordrecht, The Netherlands. pp. 631-40.

Bhojwani, S.S. and Razdan, M.K. 1983. *Plant Tissue Culture–Theory and Practice.* Elsevier, Amsterdam, The Netherlands.

Bollon, H. and Raquin, C. 1987. Haplomethods: a tool for crop improvement. In Nestec Ltd. (ed), *Nestle Research News,* Nestec Ltd., Veyvey, Switzerland. pp. 81-93.

Czygan, F.C. (ed). 1984. *Biogene Arzneistoffe.* Vieweg, Braunschweig, Germany.

DiCosmo, F. 1990. Strategies to improve yields of secondary metabolites to industrially interesting levels. In H.J.J. Nijkamp, L.H.W. van der Plas and J. van Aartrijk (eds), *Progress in Plant Cellular and Molecular Biology.* Kluwer Academic Publishers, Dordrecht, The Netherlands. pp. 754-63.

Dixon R.A. (ed) 1987. *Plant Cell Culture–A Practical Approach.* IRL Press, Oxford, Washington.

Evans, D.A. and Sharp, W.A.1986. Applications of somaclonal variation. *Biotechnology,* 4, 528-32.

Fontanel, A. and Tabata, M. 1987. Production of secondary metabolites by plant tissue and cell cultures–present aspects and prospects. In Nestec Ltd. (ed), *Nestle Research News,* Nestec Ltd., Veyvey, Switzerland. pp. 93-105.

Galneder, E. and Zenk, M.H., 1990. Enzymology of alkaloid production in plant cell cultures. In H.J.J. Nijkamp, L.H.W. van der Plas and J. van Aatrijk (eds), *Progress in plant Cellular and Molecular Biology.* Kluwer Academic Publishers, Dordrecht, The Netherlands. pp. 754- 63.

Ruyter, C.M. and Stockigt, J., 1989. Neue Naturstoffe aus pflanzlichen Zell- und Gewebekulturen. *GIT Fach.* Lab., 33, 283-93.

Spencer, A., Hamill, J.D., Reynolds, J. and Rhodes, M.J.C. 1990. Production of terpenes by transformed differentiated shoot cultures of *Mentha piperita citrata* and *M. piperita vulgaris.* In H.J.J. Nijkamp, L.H.W. van der Plas and J. van Aartrijk (eds), *Progress in Plant Cellular and Molecular Biology.* Kluwer Academic Publishers, Dordrecht, The Netherlands. pp. 619-25.

Stafford, A. and Fowler, M.W. 1991. Plant cell culture and product opportunities. *Agro-Industry* Hi-Tech, 2, 19-23.

Westphal, K. 1990. Large-scale production of new biologically active compounds in plant cell cultures. In H.J.J. Nijkamp, L.H.W. van der Plas and J. van Aartrijk (eds) , *Progress in Plant Cellular and Molecular Biology.* Kluwer Academic Publishers, Dordrecht, The Netherlands. pp. 601-09.

Wilson, P.D.G., Hilton, M.G., Meehan, P.T.H., Waspe, C.R. and Rhodes, M.J.C., 1990. The cultivation of transformed roots from laboratory to pilot plant. In H.J.J. Nijkamp, L.H.W. van der Plas and J. van Aartrijk (eds), *Progress in Plant Cellular and Molecular Biology.* Kluwer Academic Publishers, Dordrecht, The Netherlands. pp. 700-06.

Yamada, Y. and Hashimoto, T. 1990. Possibilities of improving yields of secondary metabolites in plant cell cultures. In H.J.J. Nijkamp, L.H.W. van der Plas and J. van Aartrijk (eds), *Progress in Plant Cellular and Molecular Biology.* Kluwer Academic Publishers, Dordrecht, The Netherlands. pp. 547-58.

10

SURVIVAL STRATEGIES OF BACTERIA IN THE NATURAL ENVIRONMENT AND THE SEARCH FOR NEW MICRO-ORGANISMS

F. TORRELLA
University of Murcia, Spain

The field of biotechnology is a multidisciplinary area of applied science which has received input from a variety of disciplines, namely, microbiology, biochemistry genetics, and biochemical engineering. Micro-organisms have been and still are one of the basic raw materials for biotechnology (Malik, 1989) and this fact is implicitly acknowledged in broad-scope books such as Hopwood and Chater (1989).

A knowledge of how new micro-organisms adapt to survive and thrive in their natural habitats should be of great help to groups of scientists involved in broad biotechnological innovative research projects. The study of the autoecology of known and newly isolated microbes fulfils an essential role since the extraordinary diversity of prokaryotic genomes and the adaptations to their habitats is an endless source of new products and ideas for a biotechnology team.

It is well known and accepted by classical microbiologists and modern microbial ecologists that the micro-organisms that may be recovered (isolated) from natural habitats are but a very small fraction of those observed therein. This has been one of the drawbacks to the progress of microbial ecology (Colwell *et al.*, 1989) and one may wonder if this drawback could also apply to biotechnology. We will only know the possibilities offered by organisms still not cultured to biotechnology when these micro-organisms are isolated and properly characterized in the laboratory. Proper characterization of the new isolates is essential since substantial parts of the isolated genomes may remain silent in culture owing to 'narrow-minded' cultivation techniques which fail to take into account the natural habitat parameters of where the organisms have evolved. There is still a vast amount of work to be done by 'microbe hunters', and there must be a place for them in the current panorama of the biotechnology.

As proof of the above, in recent years, the isolation of new micro-organisms from the so-called extreme environments has received much attention: see, for example, Sharp and Munster (1986), and recent publications on extremophiles. On the other hand, it is becoming more and more evident that 'normal habitats' such as marine and freshwaters, and soils, still have much to offer in terms of diversity of micro-organisms not yet isolated. This has been demonstrated by the recent work of Britschgi and Giovannoni (1991) and Giovannoni *et al.* (1990). By using molecular genetic methods to examine the taxonomical structure of natural microbial communities, the latter authors have demonstrated that there is still much to be 'fished out' from these common habitats. Luckily for the classical microbe hunter, careful and imaginative isolation techniques based on such well known strategies as enrichments, and point dilutions and plate cultures, may still offer to 'prepared eyes and mind' new and interesting microbial types, even among heterotrophic microbial forms. When dealing with isolation of new heterotrophic bacteria from waters and soils, concepts such as oligotrophs and copiotrophs, autochthonous and zymogenous, ultramicrobacteria, dwarfs, minicells, starvation-survival, etc., come immediately to the arena of the scientific discussion. For example, there is controversy concerning the existence of very small bacteria which would grow very slowly(adaptation to oligotrophy) and barely increase in size upon resuscitation and cultivation as was shown years ago through slide cultures and time-laspse photography by Torrella and Morita (1981). Some authors doubt if such a class of microbes does in fact exist, since they have not been able to isolate such forms of oligotrophic bacteria (McDonell and Hood, 1982; Tabor *et al.*, 1981; Martin and Macleod, 1984). As some authors report, the possibility exists that, after all, ultramicrobacteria do increase in size when the proper nutrients and cultural conditions are employed. Since this controversy falls into the area of new micro-organism isolation discussed above, and also touches the research area on starvation- survival, some examples from the author's work, which involves isolation and characterization of very small size aquatic bacteria (ultramicrobacteria), are relevant here (Torrella and Morita, 1981; Torrella, 1989, 1990).

As mentioned above, the ultramicrobacteria subject and the starvation-survival adaptation of aquatic microbes are much related areas of research since natural habitats are mostly oligotrophic in terms of organic matter concentration. See Morita (1988) , Ishida *et al*, (1990), and references therein. The survival of bacteria in natural waters has had to face the fact of the scarcity of energy available and the adaptations to starvation have been demonstrated to be extremely diverse, ranging from morphological changes (small cell formation), to metabolic and physiological adaptations (Kjelleberg *et al.*, 1987; Moyer and Morita, 1989; Martin *et at.*, 1989, and references therein). Since the starvation-survival state of micro-organisms in nature is most frequently the rule and not the exception, a knowledge of the adaptations of the aquatic microbes to this situation will be a sure ally to the microbe hunter. Benefit of this knowledge may be found in areas as fundamental in micro-organism isolation as in finding new strategies for successful resuscitation of dormant cells and design and/ or optimization of isolation techniques, to cite just two examples. Other aspects of the implications of the starvation–survival area of research in the biotechnology, including plasmid stability, cell absorption, cell immobilization in gels, release of engineered microbes, gene expression in nutrient deficient media, etc., should be considered.

REFERENCES

Britschgi, G.I. and Giovannoni, S.J. 1991. Phylogenetic analysis of a natural marine bacterioplankton population by RNA gene cloning and sequencing. *Appl. Environm. Microbiol.*, 57, 1707-13.

Colwell, R.R., Knight, I.T., Summerville, C., Shults, C. and Kaspar, C.W. 1989. Viable but non-culturable phenomenon in relationship to starvation/survival, 'injury' and strategies of survival of bacteria in the environment. In Hattori *et al.* (eds), *Recent Advances in Microbial Ecology.* Proc. ISME-5, Jpn Sci. Soc. Press, Tokyo, 1989.

Giovannoni, S.J., Britschgi, T.B., Moyer, C.L. and Field, K.G. 1990. Genetic diversity in Sargasso Sea bacterioplankton. Nature, 345, 60-63.

Hopwood, D.A. and Chater, K.F. (eds). 1989. *Genetics of Bacterial Diversity.* Academic Press, London.

Ishida, Y., Fukami, A., Eguchi, M. and Yoshinaga, 1. 1990. Strategies of growth of oligotrophic bacteria in the pelagic environment. In Hattori *et al.* (eds), *Recent Advances in Microbial Ecology.* Proc. ISMe-5, Jpn Sci. Soc. Press, Tokyo. pp. 89-93.

Kjelleberg, J., Hermasson, M., Marden, P. and Jones, G.W. 1987. The transient phase between growth and non-growth of heterotrophic bacteria, with emphasis on the marine environment. *Ann. Rev. Microbiol.*, 41, 25-49.

Malik, V.S. 1989. Biotechnology–the Golden Age. *Adv. Appl. Microbiol.* (Academic Press) 34, 263-306.

Martin, A., Auger, E.A., Blum, P.H. and Schultz, J.E. 1989. Genetic basis of starvation survival in non-differentiating bacteria. *Ann. Rev. Microbiol.*, 43, 293-316.

Morita, R.Y. 1988. Bioavailability of energy and its relationship to growth and starvation survival in nature. Can. J. Microbiol., 34, 436-41.

Moyer, C.L. and Morita, R.Y. 1989. Effective growth rate and starvation survival on the viability and stability of a psychophylic marine bacterium. *Appl. Environm. Microbiol.*, 55, 1122-27.

Sharp, R.J. and Munster, M.J. 1986. Biotechnological implications for micro-organisms from extreme environments. In R.A. Herbert and G.A. Codd (eds), *Microbes in extreme environments.* Academic Press, London. Ch. 9.

Torrella, F. 1989. Slow-growing minute bacteria from marine plankton: a most looked for group of micro-organisms. In Hattori *et al.* (eds), *Recent Advances in Microbial Ecology.* Proc. ISME-5, Jpn Sci. Soc. Press, Tokyo. Contrib. Paper 0-11-7. Abstr. Book.

Torrella, F. 1990. Form and function of minute prokaryotes and their role in the carbon/energy cycle in the sea. Keynote Paper in session *Cycling of Energy and Elements.* Abstr. Book 4th Europ. Marine Microbiology Symp., Kiel.

Torrella, F. and Morita, R.Y. 1981. Microcultural study of bacterial size changes and microcolony and ultramicrocolony formation by heterotrophic bacteria in sea water. *Appl. Environm. Microbiol.*, 41, 518-27.

ABBREVIATIONS

AA	anthranilic acid
AASase	anthranilate synthase
ASR	alkaloid synthesis rate
7-Aza-ind	7-aza-indole
7-Aza-trp	7-aza-tryptophan
CA	chorismic acid
C.f.	*Claviceps fusiformis*
C.p.	*Claviceps purpurea*
CM	chorismate mutase
Cp	cyclopeptine
col	cyclopenol
cyn.	cyclopenine

cyt.	cytoplasmatic
DAHP	3-deoxy-D-arabino-heptulosonic acid-7 phosphate
DAPH-Sase	DAHP synthase
DM	dry mass
ext.	external
5-F-ind	5-fluoro-indole
5-F-trp	5-fluoro-tryptophan
germ. phase	germination phase
5-OH-ind	5-hydroxy-indole
5-OH-trp	5-hydroxy-tryptophan
InGP	indole-3-glycerol-phosphate
4-Me-trp	4-methyl-tryptophan
5-Me-trp	5-methyl-tryptophan
6-Me-trp	6-methyl-tryptophan
PA	prephenic acid
P.c.	*Penicillium cyclopium*
SAM	S- adenosy1- methionine
TrpSase	tryptophan synthase
vac.	vacuolar

11

REGULATORY INTERACTIONS BETWEEN PRIMARY AND SECONDARY METABOLISM AS REGULATING PRINCIPLES IN THE PRODUCTION OF SECONDARY METABOLITES

H-P. SCHMAUDER
Friedrich-Schiller Universitat, Jena, Germany

Summary. An overview is given on known basic principles of cellular regulation principles. Along with general mechanisms, the principles are demonstrated on the example of biosynthesis of aromatic amino acids and their relation to the formation of secondary metabolites derived from these amino acids. As examples, results obtained with strains of *Claviceps* and *Penicillium cyclopium* are discussed. Strains of *Claviceps* show a negative feedback of alkaloids to biosynthesis of precursors and of alkaloids. On the other hand, alkaloids formed by *Penicillium cyclopium* stimulate the enzymes of the aromatic pathway of this fungus. This is one of the first described activations of secondary metabolites on precursor formation in micro organisms.

GENERAL PRINCIPLES OF METABOLIC REGULATION

Textbooks describe various possibilities for regulation of metabolism (see, for an example, Schlegel, 1985). The following mechanisms are known:

1. Regulation at genomic level: induction and repression.

In these areas, an activation or inhibition of the metabolism is possible by effects of substrates, enzymes, products, specific living conditions, etc. These effects are dependent on different external as well as internal signals. The reaction is possible in a co-ordinataed and sequential way.

The differential gene expression is dependent on the activities, structures, and co-operation of operators, repressors, promoters, and the structure of different operons.

Repressions are possible by end or intermediary products or as katabolite repression.

2. Regulation in the enzyme level: feedback and feedforward regulation of enzymes.

Binding to effectors of the protein, modification of proteins, isoenzymes, and different allosteric mechanisms are the main principles of these kinds of regulation.

Some exergonic or endergonic types of reactions mainly on the outer phases of proteins are also able to regulate the activities of enzymes. All these regulations are possible in a co-ordinated as well as sequential manner.

3. A third influence is given by effects of channelling, transport between the cells, between cells and medium, and within the cells and some other structural systems of the cells.

A lot of open questions are connected with the problems of interaction of primary and secondary metabolism. Are there regulatory interactions? So far, only negative effects of secondary metabolities on the regulation of precursor biosynthesis have been described in textbooks. Is this the rule or are exceptiions possible?

Regulation of Biosynthesis of Aromatic Amino Acids and of Biosynthesis of Secondary Products Derived from Intermediates or Products of the Aromatic Pathway.

In the aromatic pathway for biosynthesis of aromatic amino acids, some precursors for secondary metabolites were synthesized. From shikimic acid, chorismate, prephenate, anthranilic acid, tryptophan, tyrosine as well as phenylalanine alkaloids, phenolic compounds, flavonoids, etc., were formed. The aromatic pathway is regulated by operons as well as on the enzymatic level. The last form of regulation is summarized in a short version in Fig. 11.1.

Commonly, the four key enzymes DAHP synthase, chorismate mutase, anthranilate synthase and rarely tryptophan synthase were regulated by feedback effects of the final products of these biosynthetic pathway: the amino acids phenylalanine, tyrosine and tryptophan inhibit the DAHP synthase. The value of decreasing the activity is dependent on the isoenzymes existing and the used organisms. The chorismate mutase is influenced mostly by tyrosine and phenylalanine in a negative, and in several organisms by tryptophan in a positive, manner. For example, this enzyme isolated from *Claviceps paspali* can be used as an agent for quantification of *tryptophan* (Sprossler and Lingens, 1970a, b). Both end-products phenylalanine and tyrosine also inhibit the branching enzyme chorismate mutase, phenylalanine the prephenate dehydratase and tyrosine the prephenate dehydrogenase. The anthranilic acid synthase is inhibitable in a more or less high degree by tryptophan. Only the tryptophan synthase shows mostly no feedback regulation by one of the end-products of the aromatic pathway (see Fig. 11.1).

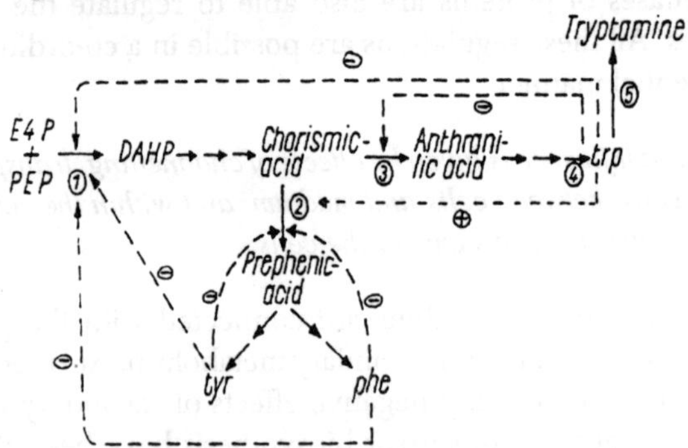

Fig. 11.1: Principles of regulation of the pathway of biosynthesis of aromatic amino acids. ⊖ : inhibition, ⊕ : activation.

Many experiments were performed using antimetabolites or antimetabolite-like substances with the intention of isolating strains with a remarkable overproduction of the aromatic amino acids as precursors of secondary product formation. But in some cases this overproduction does not contribute to a higher production rate of secondary products.

A characteristic positive example is given by Sasse *et al.* (1983a, b). By selection of wild types of *Catharanthus roseus* for a resistance to derivatives of tryptophan lines resistant to 4-Me-trp showed a higher activity of trp-decarboxylase and a higher accumulation of tryptamine. This compound is a precursor for indole alkaloid biosynthesis and shows mostly a negative feedback on the precursor formation. The described lines resistant to 4-Me-trp were able to produce higher amounts of this precursor and of the alkaloid ajmalicine in comparison with the wild-type strains.

These resistant lines also contained a higher level of tryptophan, but the feedback regulation of AASase by this amino acid was not altered in relation to the wild types!

Experiments with CLAVICEPS

Claviceps forms ergot alkaloids from trp and a mevalonic acid derivative. The alkaloids show on the one hand a negative feedback to precursor formation and the first enzyme(s) of the alkaloid formation pathway. For example, Heinstein and Floss (1976) and Cheng *et al.*, (1980) reported a negative feedback of alkaloids, e.g., elymoclavine, on the first enzyme of alkaloid formation, the dimethyl allyltryptophan synthase in cells and protoplasts of *Claviceps fusiforis*, strain SD 58. A negative interaction of elymoclavine on AASase was described by Schmauder (1975), Floss (1976), Schmauder and Groger (1973, 1976) as well as Mann and Floss (1977).

Feedback effects of TrpSase are rare. This enzyme of *Claviceps purpurea* reacts on ergot alkaloids via a (possible) repression. Fig. 11.2 shows results of experiments with two different strains of *Claviceps purpurea* able and unable to produce ergotoxins (Schmauder, 1975). Strain M 5/95 F is only a produce by parasitic cultivation on rye and is alkaloid-free by saprophytic cultivation.

Also, under saprophytic cultivation techniques, a negative response of its TrpSase on alkaloids of the ergotoxine-group is found. In this reaction, there are similarities to the saprophytic alkaloid producer Pepty 695 with one exception. Generally, in the strain Pepty 695, the TrpSase is not detectable at the 8th day.

This finding of a similar reaction of the two strains used indicated the existence of a more general consequence of this form of regulation.

Experiments with different derivatives of tryptophan with the intention to increased resistance, a higher biosynthetic and accumulation rate for tryptophan and alkaloids were performed for instance in the laboratories of Schmauder (Germany) and Pazoutova (Czechoslovakia). The main results obtained, and also reproduced in Schmauder and Groger (1976), Schmauder *et al.* (1976) and Pazoutova *et al.* (private communication, 1991) will be summarized briefly. Table 11.1 shows a general overview of results obtained in the author's laboratories. It is interesting that only 4-Me-trp and the 5-fluorinated trp and indole show an effect.

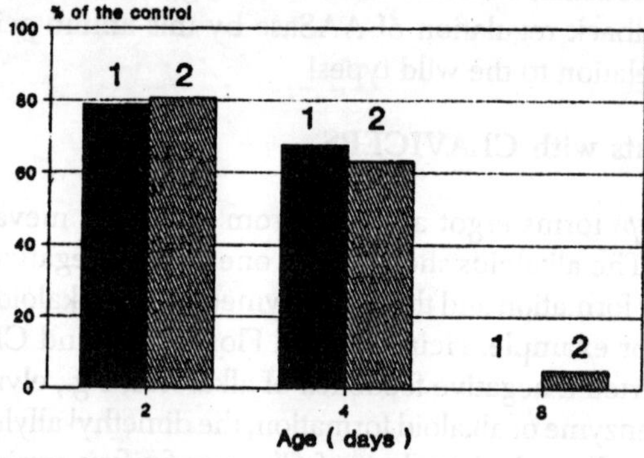

Fig. 11.2. Influence of ergotoxine alkaloids on TrpSase in cultures of Claviceps purpurea. Strains used: Pepty 695, able to form ergotoxine alkaloids, and M 5/95 F, wild type strain, isolated from sclerotia, unable to form alkaloids in saprophytic culture. The cultures were supplemented with crude alkaloid mixture isolated from cultures of Pepty 695 six hours after inoculation. The final concentration of alkaloids was 1 mM. All values are given in % of the untreated control. Generally TrpSase is not detectable in 8 day old cultures of Pepty 695 (see also Schmauder, 1975).

Cultivation with 5-Me-trp was possible in a nearly saturated solution without significant influence. Probably there is a limitation to the uptake of this substance into the cells. The effects of the two aza-derivatives are only based on an influence on the activity of TrpSase. The alkaloid formation is nearly normal and the effect on growth and alkaloid formation is normalizable by trp.

A drastic effect on growth and alkaloid formation of *Claviceps* show the 5-fluorinated derivatives of indole and tryptophan. Very small amounts (5-100 ug/ml culture liquid) inhibit growth and alkaloid formation. The effect depends on the compound, concentration and time of addition. 5-fluoro-indole induces the inhibition at lower concentrations than 5 fluoro-tryptophan.

Both compounds induce a response only after addition to the cultures in the first 18 h of cultivation. Later addition is without any effects. The inbibition of growth is reversible by adding the same amount of trp. This reversion is not bound to the time of addition. But in all cases of addition of trp, only the growth inhibition is reversible. The alkaloid formation is not normalized by the amino acid. The activities of the enzymes of the aromatic pathway were also influenced by the fluorinated compounds: there is a remarkable increase in the activities of enzymes in these cultures treated with the fluorinated compounds alone. A simultaneous feeding with trp or with trp alone did not change significantly the enzyme activities (Schmauder *et al.*, 1976).

TABLE 11.1
EFFECT OF DIFFERENT ANTIMETABOLITES OF TRYPOTPHAN ON GROWTH AND ALKALOID FORMATION OF *CALVICEPS* (SCHMAUDER AND GROGER, 1976; SCHMAUDER *ET AL.*, 1976; SCHMAUDER, UNPUBLISHED).

Strain	*Antimetabolite compound*	*concentration*	*Effect*
C.p.	4-Me-trp	0.01mM	strong inhibition of alkaloid formation; competition to binding sites of enzymes
C.p. C.f.	5-Me-trp	up to mM	no effect detectable on growth and alkaloid formation
C.p.	6-Me-trp	up to 1 mM	no effect detectable on growth and alkaloid formation
C.f.	7-Aza-trp 7-Aza-ind	0.001-1 m	Mslight inhibition of growth and alkaloid formation, regenerable by addition tryptophan; inhibition of tryptophan synthase
C.P. C.f.	5-OH-trp	up to 1 mM	no effect detectable on growth and alkaloid formation
C.p. C.f.	5-F- trp 5-F-ind	5-100 Mg.ml	strong inhibition of growth and alkaloid formation, growth effects will be compensated by addition of the same amounts of trypotphan; alkaloid formation will not be normalized; effects are strongly time-dependent

Figure 11.3 summarizes the results of feeding experiments of the ergotoxine-forming strain C.p. Pepty 695. In a series of experiments of cultivation with replacements of washed, pregrown, alkaloid synthesizing mycelium on the one hand, and a normal inoculation (by using a ratio of about 1:10) on the other hand, in new media, with and without antimetabolite growth and alkaloid formation, were followed up. It is concluded that there is a longer acting effect of the fluorinated derivatives because a regeneration of alkaloid formation to the level of the control needs at least 2-3 cultivation periods without the antimetabolites (Schmauder, unpublished). The basis of this action of fluoro-derivatives on the strain *Claviceps purpurea Pepty* 695 used in our experiments is not yet known (possibly at the genomic level?).

EXPERIMENTS WITH *PENICILLIUM CYCLOPIUM*

Schmauder and Roos (1987) and Roos and Schmauder (1989) have investigated the interaction between primary and secondary metabolism in *Penicillium cyclopium*. The alkaloids formed in emerged cultures of this fungus, e.g., cyclopeptine, cyclopenine, and cyclopenol, are biosynthetically derived from methionine, phenylalanine and anthranilic acid (Fig. 11.4). In this case, the regulation between the different areas of biosynthesis is of interest.

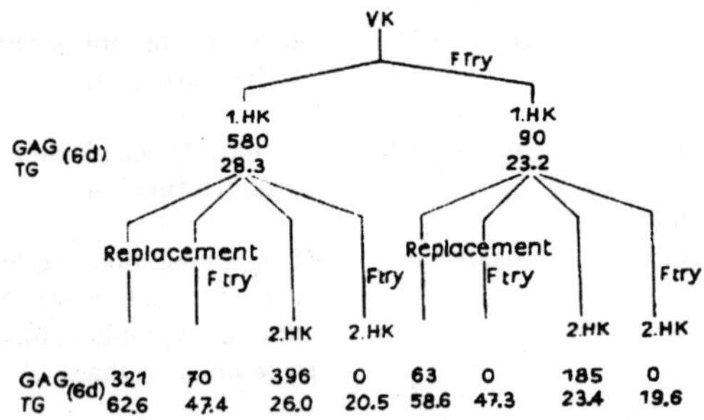

Figure 11.3: (Schmauder, unpublished.) Influence of 5-fluoro-tryptophan on cultures of Claviceps purpurea, strain Pepty 695. The cultures were supplemented with the antimetabolite of tryptophane as follows. (a) Replacement of tho grown mycelium in new culture liquid, supplemented with or without fluoro-tryptophan (b) inoculated in culture liquid supplemented with or without fluoro-tryptophan (inoculation rate:11.:10).

VK	preculture (NL 720) Schmauder and Groger 1973)
HK	main culture (NL 720)
GAG	total alkaloid content (μg/ml)
Ftry	5-fluoro-tryptophan.

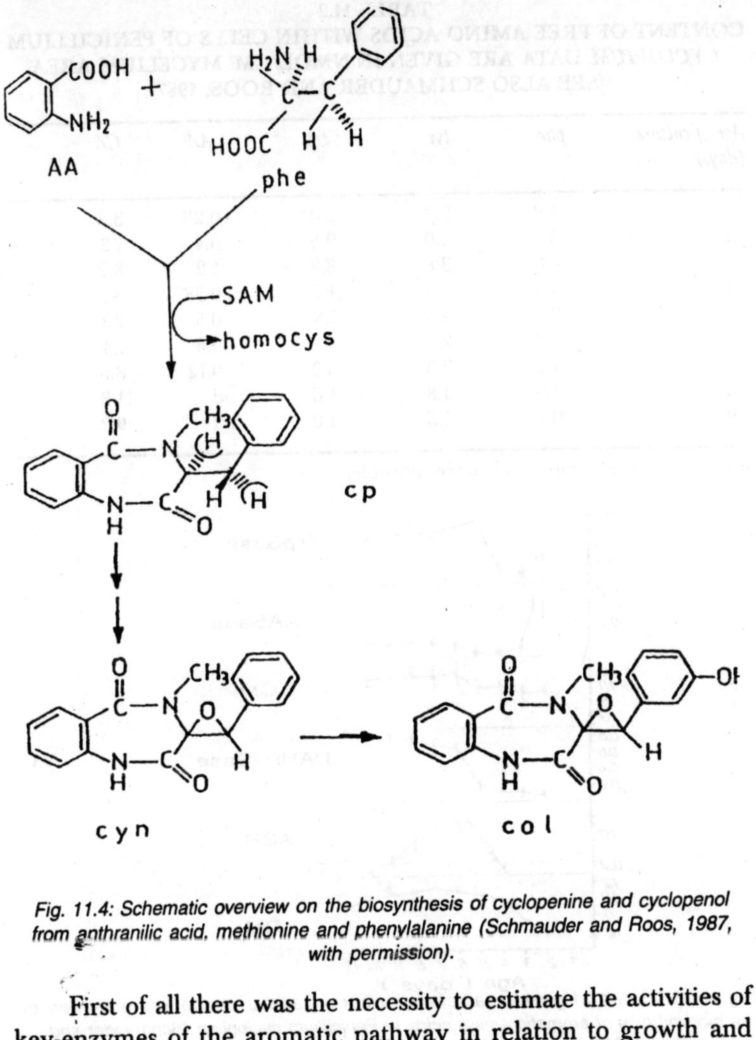

Fig. 11.4: Schematic overview on the biosynthesis of cyclopenine and cyclopenol from anthranilic acid, methionine and phenylalanine (Schmauder and Roos, 1987, with permission).

First of all there was the necessity to estimate the activities of key-enzymes of the aromatic pathway in relation to growth and alkaloid formation. Figure 11.5 shows the surprising finding that apparently in *Penicillium cyclopium* the activities of DAHP-Sase, CM, and AASase are increasing in the range of highest alkaloid formation rate. In these parts of cultivation, the intracellular contents of the aromatic amino acids phe, tyr, trp, and also of chorismic acid as a branching point in the aromatic pathway, and anthranilic acid, showed a maximum. Because the anthranilic acid is a strong cytotoxic substance, the level of this compound is lower than of the other compounds studied (Table 11.2).

TABLE 11.2
CONTENT OF FREE AMINO ACIDS WITHIN CELLS OF PENICILLIUM *CYCLOPIUM*. DATA ARE GIVEN IN NMOL/CM² MYCELIUM AREA (SEE ALSO SCHMAUDER AND ROOS, 1987).

Age of culture (days)	*phe*	*tyr*	*trp*	*AA*	*CA*
1	1.9	2.2	2.0	0.28	3.2
2	3.4	3.0	9.5	0.4	7.2
3	4.0	2.8	8.8	1.2	8.2
4	5.3	4.5	11.2	0.38	8.0
5	2.2	2.5	7.8	0.5	7.8
6	1.4	2.0	5.8	0.8	7.4
7	1.2	2.0	6.2	0.12	8.0
8	1.0	1.8	4.0	nd	11.8
9	0.8	1.5	4.0	nd	9.2

nd: not detectable, only traces possible.

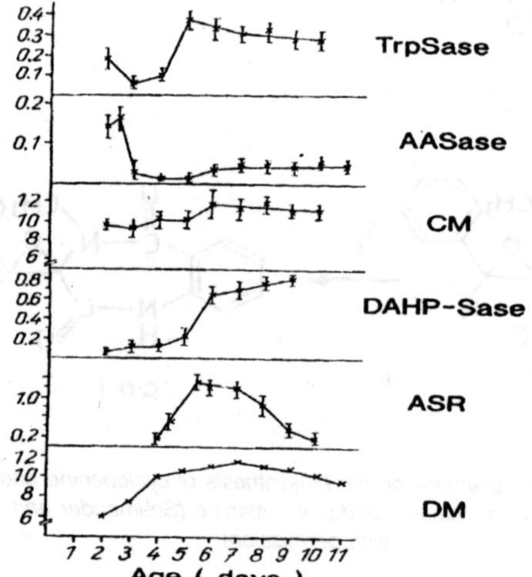

Fig. 11.5: Rates of growth, alkaloid formation and of some key-enzymes of the biosynthesis of aromatic amino acids in Penicillium cyclopium (Schmauder and Roos, 1987, with permission). The rates are given in the following units:

enzyme activities *nkat/10 mg of proteins*

alkaloid synthesis *μ mol alkaloid/cm² x 24 h*

dry mass *mg TM/cm² mycelium area.*

During these experiments it was found that the main key-enzymes DAHP-Sase, CM and AASase showed an increase in activity after starting the alkaloid synthesis. In most other known systems the opposite finding is described (Hertel *et al.*, 1988;

Schmauder *et al.*, 1984, 1985; etc.). By following this effect an activation of the precursor formation was observed: after starting the alkaloid synthesis the activities of these three enzymes are increasing. This increase is not only observable in cultures fed with alkaloids (see also Fig. 11.5) but also in cell-free *in vitro* systems. Table 11.3 summarizes the results with these cell-free extracts: with the exception of TrpSase, all studied enzymes show an increased activation by alkaloids. The highest intensity in the reaction is shown by the anthranilic acid synthase. This high activation pattern is necessary because anthranilic acid is within the cells in a minimum concentration as a cytotoxic compound.

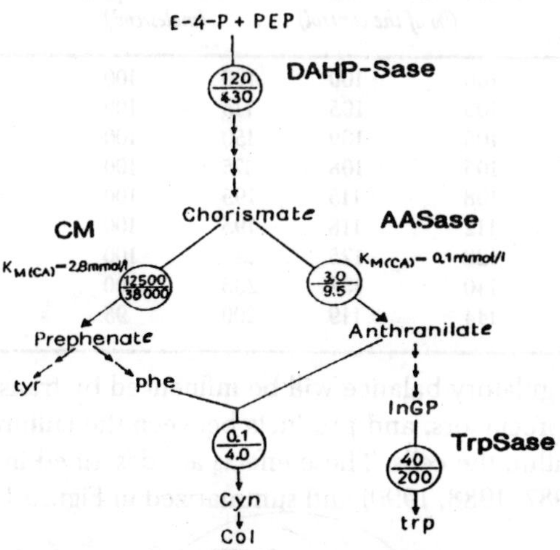

Fig. 11.6: Comparison of maximum in vitro turnover rates of important key-enzymes of the biosynthesis of aromatic amino acids with in vitro rates of synathesis of cyclopenine and cyclopenole in emerged cultures of Penicillium cyclopium. The number in circles contain the valuable turnover rates: (a) upper values from 3rd day of cultivation, before alkaloid synthesis started; (b) lower values from 9th day of cultivation during alkaloid formation (Schmauder and Roos, 1987, with permission).

The flux rates through the aromatic pathway as well as the alkaloid synthesis is strictly regulated by the culture age and alkaloid content. In the first phase of cultivation, a slow flux rate was measured followed by an increase after starting the alkaloid biosynthesis. In this later phase, mainly AASase shows the highest growth rate in its activity. It should be noted that the KM-value of chorismate mutase is much higher than the value of AASase. Under these conditions, the following regulationary cascade will be active: if there are lower concentrations of chorismic acid within the cells, the lower KM-value enables a higher activity of AASase and the CM acts only under optimal conditions after reaching higher chorismate concentrations. These regulationary situations allow a balance between primary metabolism, precursor biosynthesis and alkaloid formation (see also Schmauder and Roos, 1987; Roos and Schmauder, 1989).

TABLE: 11.3
STIMULATION OF *IN VITRO* ACTIVTIES OF KEY-ENZYMES OF BIOSYNTHESIS OF AROMATIC AMINO ACIDS IN *PENICILLIUM CYCLOPIUM* CELL-FREE EXTRACTS (SEE ALSO ROOS AND SCHMAUDER, 1989).

Age of culture (days)	*DAHP-Sase*	*CM*	*AASase*	*TrpSase*	*Alkaloid content*
	(% of the control)				*(moles/cm²)*
2	100	100	100	100	0
3	100	105	115	100	0.05
4	105	109	150	100	0.2
5	105	108	175	100	0.38
6	108	115	195	100	0.68
7	112	118	195	100	0.82
8	128	125	215	100	0.92
9	140	138	235	100	1.05
10	144	119	200	95	1.09

The regulatory balance will be influenced by transportation of substrates, precursors, and products between the culture liquid, the cells and within the cells. These effects are described in more detail by Roos (1987, 1988, 1990) and summarized in Figure 11.7.

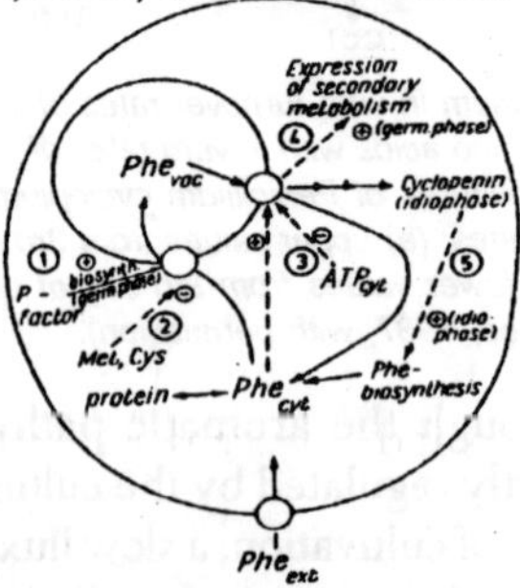

Fig. 11.7: Overview of regulatory mechanisms acting at the level of transport and channelling of the alkaloid precursor phenylalanine in Penicillium cyclopium (from Roos, 1990, with permission).

(1) under the influence of P-factor the biosynthesis of vacuolar phenylalanine carriers is stimulated
(2) above a threshold concentration, cellular methionine and cysteine inactivate vacuolar phenylalanine carriers

(3) distinct concentrations of cellular ATP inhibit the efflux from the vacuole; high levels of cytosolic amino acids stimulate efflux in the presence of sufficient ATP
(4) the vacuolar phenylalanine pool is most probably involved in triggering the expression of alkaloid metabolism
(5) in the idiophase, cyclopenine stimulates enzymes involved in the biosynathesis of phenylalanine.

CONCLUSIONS

After first finding of a positive feedback of alkaloids on primary metabolism described by Schmauder and Roos (1987) and Roos and Schmauder (1989) and discussed in short from above, the following regulatory interactions between primary and secondary metabolism are possible:

1. There is mainly a negative feedback acting. These effects are described for instance for *Claviceps* (*e.g.*, Schmauder, 1975; Floss, 1976; Cheng *et al.*, 1980), *Ruta graveolens* (*e.g.*, Hertel *et al.*, 1988), *Nigella damascena* and *Cinchona succirubra* (Schmauder *et al.*, 1984, 1985).

Higher activities of secondary products repress the enzymes of the aromatic pathway and the formation of precursors for secondary metabolite formation. In the phases of culture with a higher content on secondary products often no activities of enzymes of the aromatic pathway are detectable, *e.g., Claviceps* (Schmauder and Groger, 1973), *Cinchona succirubra* (Schmauder *et al.*, 1985).

2. Under specific conditions, an activation of precursor formation described above is realized, to allow a sufficient pool of precursors of this specific syntheses. The biological role of this type of regulation is unclear.

3. An overlapping function could be that of 'general regulation' (*e.g.*, of amino acid biosynthesis), channelling, as well as transport between cells, the cells and the nutrient broths and within the cells, *e.g.*, described by Roos (1987, 1988, 1990), Schmauder (1987) and Roos and Roos and Schmauder (1989).

REFERENCES

Cheng, L.J., Robbers, J.E. and Floss, H.G. 1980. End-product regulation of ergot alkaloid formation in intact cells and protoplasts of Claviceps species, strain SD 58. *J. Nat. Prod.*, 43, 329-39.

Floss, H. 1976. Regulation of alkaloid synthesis in the ergot fungus. *Microbiol.*, 553-54.

Heinstein, P. and Floss, H.G. 1976. Feedback inhibition and end-product repression in secondary metabolism. *Nova Acta Leopoldina*, Suppl. No. 7, 299-310.

Hertel, S.C., Schmauder, H. P., Hieke, M. and Groger, D. 1988. Alkaloidbildung und Biosynthese aromatischer Aminosauren in Zellkulturen von Ruta graveolens L. Biochem. Physiol. Pflanzen, 183, 425-37.

Mann, D.F. and Floss, H.G. 1977. Purification and properties of anthranilate synthetase from the ergot fungus, *Claviceps* species, strain SD 58. *Lloydia*, 40, 136-45.

Roos, W. 1987. Regulationsprinzipien im mikrobiellen Stoffwechsel- die besondere Rolle der zellularen Raumstruktur. *Wiss. Z. Univ. Halle*, 36, 65-76.

Roos, W. 1988. Compartmentation and transport of precursors in the regulation of secondary metabolism in eukaryotic micro-organisms. *EFB, Proc. 4th Int, Workshop*, OMIKK, Budapest. pp. 157-68.

Roos, W. 1990. Benzodiazepine alkaloids. *The Alkaloids,* 39, 63-97.

Roos, W. and Schmauder, H. P. 1989. Positive feedback effect of benzodiazepine alkaloids on enzymes of the aromatic pathway. *FEMS Microbiol. Lett.*, 59 27-30.

Sasse, F., Buchholz, M. and Berlin, J. 1983a. Site of action growth inhibitory tryptophan analogues in *Catharanthus roseus* cell suspension cultures. *Z. Naturforsch.,* 38c, 910-15.

Sasse, F., Buchholz, M. and Berlin, J. 1983b. Selection of cell lines of *Catharanthus roseus* with increased tryptophan decarboxylase activity. Z. Naturforsch., 38c, 916-22.

Schlegel, H.G. 1985. *Allgemeine Mikrobiologie.* Georg Verlag, Stuttgart, New York.

Schmauder, H. P. and Groger, D. 1973. Untersuchungen zur Aromatensynthese und Alkaloidbildung in *Claviceps-Arten. Biochem. Physiol. Pflanzen*, 164, 41-57.

Schmauder, H. P. 1975. *Beitrage zur Enzymologie und Mutation von Claviceps.* Dissert. B, AdW der DDR.

Schmauder, H. P. and Groger., D. 1976. Vergleichende Untersuchungen zur Tryptophan-Biosynthese bei *Claviceps-Arten. Biochem. Physiol. pflanzen*, 169, 201-05.

Schmauder, H. P., Gerullis, C. and Groger, D. 1976. Wirkung von 5-Fluorindol und 5-Fluor-D,L-tryptophan auf Wachstum und Alkaloidsynthese in *Claviceps. Biochem. Physiol. Pflanzen*, 170, 201-10.

Schmauder, H. P., Dobel, P. and Groger, D. 1984. Charakterisierung von *Nigella damascena*-Zellkulturen in Airlift Fermentoren. *Biochem. Physiol. Pflanzen*, 179, 611-22.

Schmauder, H. P., Groger, D., Koblitz, H. and Koblitz, D. 1985. Shikimate pathway activity in shake and fermenter cultures of *Cinchona* species. *Plant Cell Rep.*, 4, 233-36.

Schmauder, H. P. and Roos, W. 1987. Regulation von Schlussel-enzymen der Biosynthese aromatischer Aminosauren in Alkaloidbildenden Kulturen von *Penicillium cyclopium WJ. Basic Microbiol.*, 27, 583-94.

Sprossler, B. and Lingens, F. 1970a. Chorismat-Mutase aus *Claviceps*, I. Eigenschaft der Chorismat-Mutase aus verschiedenen *Claviceps-Stammen*. *Z. Physiol. Chem.*, 351, 448-58.

Sprossler, B. and Lingens, F. 1970b. Chorismat-Mutase aus *Claviceps*, II. Aktivierung und Hemmung durch Analoge der Effektoren Tryptophan, Phenylalanin und Tyrosin. Z. Physiol. *Chemi.*, 351, 967-74.

12

CHARACTERIZATION AND IDENTIFICATION OF BACTERIA

K.D. JAHNKE
DSM—German Collection of Micro-organisms and Cell Cultures, Braunschweig, Germany

Summary. Biotechnology needs named organisms with defined properties to produce substances or to transform chemicals by biological processes. Knowing the characteristics of a production strain is essential for strain improvement and process control; knowing the systematic name of an organism is the key to a wealth of useful ecological, medical, genetical, and biochemical data in the literature.

Systematics (sometimes used synonymously with classification or taxonomy) is the scientific study of the diversity of organisms and their relationships to one another. It includes three main topics:

1. classification, the ordering of organisms into groups
2. nomenclature, the assignment of correct scientific names to organisms as outlined by the International Committee of Systematic Bacteriology
3. identification, assignment of an unidentified organism to a particular group of organisms derived from classification and bearing a name given by the rules of nomenclature.

Identification or diagnosis is a comparative process by which an unknown organism is compared to known organisms. Therefore an unknown strain can only be identified if classification on the bacterial species in question has been carried out *beforehand.* It may well happen, as is often the case, that an unknown organism is isolated which cannot be identified with the presently known databases. After all, only approximately 3200 different bacterial species are known today. This is estimated to be just a very small fraction of the existing species.

INTRODUCTION

Why is it Important to Identify Bacteria?

You may ask yourself why it is necessary to go through considerable labour to describe an organism, compare it to other organisms and give it a name and an identity. In biotechnology, for example, would it not be sufficient just to know that your production organism is doing what it is expected to do?

The answer is no. If you want to have access to published information on your organism you need to know its name. If you want to save time and money it is wise practice to identify your production strain. Owing to possible bio-hazards in some countries of the world, for example, Germany, it is illegal to produce something biotechnologically with an unidentified organism.

The name of your organism is the product of the identification process. It is the key to useful ecological, medical, genetical and biochemical data.

The Trinity of Taxonomy

Systematics is the scientific study of the diversity of organisms and their relationships to one another (Claus, 1981). It includes three main topics which are sometimes called the 'trinity of taxonomy')

1. classification, the ordering of organisms into groups
2. nomenclature, the assignment of correct scientific names to organisms according to the International Code of Nomenclature (Lapage *et al.*, 1975)
3. identification, assignment of an unidentified organism to a particular group of organisms derived from classification and bearing a name given by the rules of nomenclature.

Taxonomy is the science of classification but sometimes used synonymously with the expressions classification or systematics.

The basis for classification and identification is a thorough characterization of the organism. Characterization tests range from morphological descriptions to biochemical tests for the detection of metabolic products or enzymes to highly specialized techniques such as lipid analysis, cell wall analysis, or DNA analysis.

TABLE: 12.1
SELECTED TESTS FOR THE IDENTIFICATION OF BACTERIA (ADAPTED FROM CLAUS, 1981).

Characters				
Morphological and staining	*Cultural*	*Physiological*	*Biochemical*	*Others*
Shape	Pigmentn	Phototrophic	Dissimilation	
Arrangement		Chemoheterotrophic	of sugars	
Irregular forms		Chemoheterotrophic	-oxidative	
Spores		Aerobic growth	-fermentative	
Sheaths		Anaerobic	-inactive	
Buds, appendages		Optimum temperature		
prosthecae, stalks,				
holdfasts				
Motility				
by swimming				
by gliding				
Gram reaction				

The most important classes of characters are:

morphological, cultural, physiological, nutritional, biochemical, ecological, serological, general chemotaxonomical, genetical, cell components, phage susceptibility.

Table 12.1 is an overview on the tests used in the characterization of bacteria.

Characters					
Morphological and staining	*Cultural*	*Physiological*	*Biochemical*	*Others and*	*Staining*
Acid fastness	Colonies	Temp. range	Catalase	Tween 80	Pigements
Cell inclusns.	shapes	pH range	Oxidase	Aesculin	mol%-G+H
Flagella	size	Nutrition	Gas from glucose	Utilizn. of	DNA/DNA
number	elevatn.	Need for	Acid from sugars	citrate	homology
insertion	edge	growth factors	ONPG	propinate	Cell wall comp.
Size	surface	Gaseous needs	Nitrite	malonate	Fatty acids
Ends	opacity	Heat resist.	Denitrifi-	acetate etc.	Lipids

(cont.)......

Change during	consist-	Inhibition	cation		
growth phase	ency	by	Voges-	Haemolysis	Menaquins.
	emulsif-	antibiotics	Proskauer		
	iability		Methyl red	Ferment	
Ubiquinones					
	Growth	selective	ADH	action	
	in broth	media		products	Protein
& others	surface	NaCl	ODC	& others	pattern
	turbid		LDC		Serelogy
	ity		Indol		Phase susce
			Hydrogen sulphide		ribosomal
			Urease		RNA
			Lecithinase		sequences
			Phenylalanin-		& others
			deaminase		
			Starch		
			Caselin		
			Gelatin		
			DNA		

IDENTIFICATION AND CLASSIFICATION

What is the Difference?

In principal we have to distinguish between two kinds of characterizations:

1. characterization of a bacterial strain for the purpose of classification, for example to describe a new species
2. characterization of a bacterial strain for the purpose of identification.

For the classification, strains of micro-organisms should be extensively characterized. In order to obtain enough discriminative power, determination of at least 50 and preferably several hundred characters of an organism are desirable. If you consider that the molecular weight of most bacterial DNAs varies between 1×10^6 base pairs for the small genomes of mycoplasmas and 12×10^6 nuclear base pairs for the huge genomes of certain Clostridia, you may expect 1300 to 6000 average-sized genes. So even a set of perhaps 300 tests would cover only between 5 and 20% of the genetical information of the organism.

For the purpose of identification, however, the number of tests applied should be kept to a minimum. Usually, you take only the 'key characters' which have been found to be significant for distinguishing different species. In addition, with respect to time and effort you should concentrate on simple methods which can be carried out without special equipment.

Practical Considerations

A very important step before attempting any identification is purification of the organism. This is accomplished by repeated streaking on agar or by serial dilutions. Generally, streaking is carried out by the 'three streak method' from liquid or solid media, or by the 'quadrant dilution technique' or any other techniques that will separate bacterial cells on the agar surface. It should be remembered that even if you started from a single colony and streaked out onto non-selective media, you cannot be absolutely sure that you have a pure culture. Especially with slimy cultures the procedure has to be repeated two or three times.

Standardization of Methods

Another important principle to be considered is the standardization of the methods applied for the characterization. Only if the tests are standardized is it possible to compare results described for given species. Many features used in bacterial characterization may be affected by cultural conditions. Therefore, an important general principle of characterization tests is that the culturing technique used must also be defined carefully.

Getting Started:

Finding the Correct Bacterial Group

Finding the right and shortest way to the identity of an organism depends very much on the type of organism under investigation and the personal experience of the experimenter. No definite rules can be given.

The following principles are general guidelines and were taken in large parts from the introductory article of Cowan and Liston (1974) in *Bergey's Manual of Determinative Bacteriology*, 8th edition:

- make sure that you have a pure culture, purify by repeated streaking and incubation on complex agar media such as CASO

- establish whether the organism is phototrophic, chemotrophic, or chemoheterotrophic
- examine living cells by microscopy. Look for motility, spores, sheaths, holdfasts, stalks, or cellular appendages (= prosthecae). It should be noted that some strains sporolate poorly and need to be inoculated to special media
- carry out the Gram stain, examine in bright field microscope
- look for motility of actively growing cells; do not examine old stationary phase cultures. If necessary, stain flagella and examine
- look for unusual macroscopic characters of colony form, shape and colour
- test culture for oxygen requirement (is the organism strictly aerobic or anaerobic, facultatively anaerobic, or aerotolerant?)
- if the organism is chemoheterotrophic, test oxidative or fermentative metabolism on simple sugars.

Using the information gathered, it should be possible to enter one of the Sections of *Bergey's Manual of Determinative Bacteriology*, Vols. 1 to 4 (Krieg and Holt, 1984; Sheath *et at., 1986; Stanley et al.*, 1989; Williams *et al.,* 1989; of Table 12.2. Continue by selecting additional tests as recommended by the manual or if the group is clear and well known (e.g., Enterobacteriaceae) use commercial test systems such as API (use related section) to continue. An example for the determination of the bacterial group is given in Fig. 12.1.

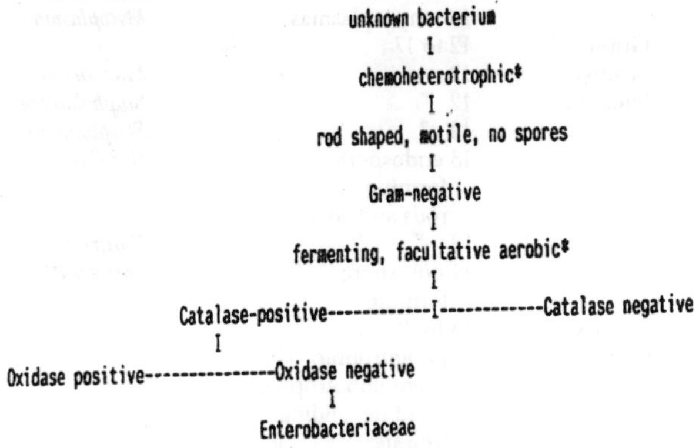

Fig. 12.2: Example for the identification of a bacterial group.

Table 12.2 has been included in this text to give a general impression of the most important groups of bacteria described in *Bergey's Manual.* Volume 1 contains mainly the Gram-negative bacteria

(Sections 1-11), Volume 2 mainly describes mainly the Gram-positive bacteria (Sections 12-17). Volume 3 deals with unusual bacteria such as phototrophic, chemolithotrophic, fruiting, gliding, and budding bacteria as well as the archaeobacteria. Volume 4, finally, deals exclusively with the actinomycetes, streptomycetes and related bacterial groups. It should be remembered that even in the latest edition of *Bergey's Manual* the grouping of the bacteria has been chosen for practical reasons of identification. They do not necessarily reflect their natural (phylogenetic) relationships.

TABLE: 12.2
LIST OF SOME IMPORTANT BACTERIAL GENERA AS DESCRIBED IN BERGEY'S MANUAL OF DETERMINATIVE BACTERIOLOGY, VOLUMES 1 TO 4.

Volume	*General bacterial group*	*Sections/ vernacular name*	*Examples of important genera*
1	Gram-negative bacteria	1 to 11	
		4 aerobic rods	*Pseudomonas*
		4 " "	*Xanthomonas*
		4 " "	*Azotobacter*
		4 " "	*Rhizonium*
		4 " "	*Agrobacterium*
		4 " "	*Alcaligenes*
		5 facultative anaerobic rods	*Escherichia*
		5 " "	*Salmonella*
		5 " "	*Klebsiella*
		5 " "	*Enterobacter*
		5 " "	*Aeromonas*
		10 mycoplasmas	*Mycoplasma*
2	Gram-positive bacteria	12 to 17	
		12 cocci	*Micrococcus*
		12 "	*Staphylococcus*
		12 "	*Streptococcus*
		13 endospore forming rods and cocci	*Bacillus*
		13 " "	*Clostridius*
		14 non-spore forming	*Lactobacillus*
2	Diverse bacteria	18 to 25 - phototrophic - chemolithotrophic - budding, gliding - fruiting - archaeobacteria	
4	Actinomycetes coryneforme bacteria	26 to 33	
		26 nocardioforme	*Nocardia*
		29 strepto-mycetes	*Streptomyces*

Miniaturized Commercial Test Systems

Although attempts were made to standardize tests for bacterial identification, no agreement has been reached on an international basis. The use of commercially available multitest systems may contribute to standardization owing to the high degree of quality control those test systems have passed, and because defined directions on their use are given by the manufacturer. So far, no commercial system has been universally accepted by microbiological or medical laboratories. In the following paragraphs, some more commonly used systems are briefly described.

The Enterotube System

Enterotubes and other similar systems are manufactured by Hoffmann LaRoche (Basel, Switzerland). Enterotube has only a limited applicability to Enterobacteria, owing to the low number of tests. Hydrated agar test media are contained inside a test tube, which is divided into 12 compartments. The tube is pierced from end to end by a straight inoculation needle, whose ends are protected by sterile caps. For inoculation one of the caps is removed, a bacterial colony is picked up and the needle is pulled through all test chambers. Test results are indicated by colour reactions. Results are evaluated by comparison with a catalogue. Dehydration problems limit the storage time of the test tubes.

The API System

This system manufactured by API bioMerieux (Montalieu Vercieu, France) is probably the most widely used in Europe. Here, the dried substrates are contained in shoe-shaped minitubes which are attached to strips. Bacterial suspensions are pipetted into the mini-tubes dissolving the substrates and the strips are then placed into covered incubation chambers which prevent the tests from drying up. For different groups of bacteria different systems exist and exact directions on their use are given.

The test results indicated by colour reactions are compared to the results in a catalogue or are evaluated with the aid of a computer. Depending on the system used, the interpretation of the results requires some skill. Accuracy is enhanced with practice and by examining reference strains with known characteristics.

Modern Methods of Chemotaxonomy and Molecular Biology

Chemotaxonomy has developed into a vast field of growing importance in bacterial taxonomy. Within this short paper only a limited number of methods can be briefly described.

Chemotaxonomy or chemosystematics is defined as the study of chemical variation in living organisms for the use in the classification and identification (Austin and Priest, 1986). With the development of modern techniques in analytical chemistry and molecular biology, chemosystematics has developed into an important supplement to traditional microbiological tests. Cell components such as chromosomal DNA, ribosomal RNA, proteins, cell wall components, membrane lipids, metabolic and products and even complete cells are the preferred objects of chemosystematics (Table 12.3).

For some groups of bacteria such as the actinomycetes, analysis of cell wall and lipids is indispensable for identification. Nucleic acid analysis, on the other hand, has been proven to be a powerful tool for the characterization of all bacteria.

One of the problems in chemosystematics is the influence of the environment on the chemical composition of bacteria. Depending on temperature, medium conditions and oxygen supply the overall cell composition of bacteria may vary considerably. Therefore it is important to standardize growth conditions when comparing bacteria on the basis of chemical components.

This, however, may be impossible, when physiologically diverse organisms, for example, thermophiles ('heat loving') psychrophiles ('cold loving') are to be compared.

Of 'all cell components used in chemosystematics only the composition of nucleic acids, namely, deoxyribonucleic acids (DNA) and ribonucleic acids (RNA), are unaffected by cultural conditions. Moreover, these 'informational molecules' are universally distributed among all organisms, which makes them ideal objects for chemosystematics and phylogenetic studies.

Analysis of Complete Cells

Bacteria may be thermally degraded in an inert atmosphere and the (unknown) break down products separated by gas chromatography and mass spectrometry (GC-MS; Gutteridge *et al.*, 1985). The resulting complex chemical profile serves as a fingerprint of the strain. Problems concerning reproducibility and the high cost of the analytical equipment have so far prevented the general success of the method.

TABLE: 12.3
MAJOR METHODS IN CHEMOTAXONOMY
(ADAPTED FROM AUSTIN AND PRIEST, 1986)

Cell component	*Method*	*Useful for the delimitation of*
Chromosomal DNA	Base composition (%GC)	Genus
	DNA:DNA reassociation	Species
	DNA fingerprinting	Species and subspecies
Ribosomal RNA	Nucleotide sequence	
	Sequence catalogue	Genus and above
	DNA:rRNA hydridization	
Proteins	Amino acid sequence	Genus and above
	Serological comparisons	
	Electrophoretic patterns	Species and Genus
	Enzyme pattern	
Cell walls	Peptidoglycan structure	
	Polysaccharides	Species and genus
	Teichoic acids	
Membrane	Fatty acids	
lipids	Polar lipids	Species and genus
	Mycolic acids	
	Isoprenoid quinones	
Metabolic products	Fatty acids	Species and genus
Complete cell	Pyrolysis/gas liquid	
	chromatography	Species and
	Pyrolysis/mass spectrom.	subspecies
	FI-IR spectroscopy	Species
	('spectral fingerprint')	

An interesting alternative to cell pyrolysis may be FT-IR spectroscopy of whole bacterial cells (Naumann *et al.*, 1991; Helm *et al.,* 1991). Infrared signals of whole bacterial cells produce highly specific fingerprint patterns in Fourier-transform infrared spectroscopy. The technique is ideal for rapid differentiation, classification and identification of bacteria. This 'non-destructive' method has produced some very promising results but its universal applicability and reliability remains to be proven.

Protein Analysis

Comparison of proteins for taxonomic purposes can be carried out either by serological methods, or by protein sequencing or, simply, by comparing total cellular protein.

Electrophoresis of whole cell proteins in polyacrylamide gels (PAGE) produces patterns which can be used as fingerprints for comparison and identification of bacteria (Jackman, 1985). SDS PAGE has found wider application. Densitometer readings of these fingerprints can be stored in the computer and can be analyzed by numerical methods.

Cell Wall Analysis

In the classical Gram-stain introduced in 1984, Gram-positive bacteria appear dark blue violet and Gram-negative bacteria will appear pink.

The reaction correlates with the structure of the cell wall. The thick cell walls of Gram-positive bacteria consist of approx. 50 layers peptidoglycan which prevents the extraction of the crystal violet/ iodine stain complex, whereas truly Gram-negative bacteria only have a thin peptidoglycan layer with a hydrophilic environment, which does not withhold the stain complex upon treatment with organic solvents.

The cell walls of most bacteria contain the polymer murein, also called peptidoglycan. (It consists of chains of the amino sugar N-acetyl-glucosamine and N-acetyl-muraminic acid, which are cross linked by peptide bridges of amino acids.

The sugar backbone is very homogeneous between different species but the tetrapeptide chain and the interpeptide bridge vary. This is used for classification especially for Gram-positive bacteria (Schleifer and Kandler, 1972).

Data on the cell wall composition are obtained from partial or complete cell wall hydrolystaes and subsequent paper or thin layer chromatography.

Lipid Analysis

The fatty acids between 0 and 20 carbons in length have been used extensively to characterize and identify bacteria, especially those belonging to the group of non-fermenting Gram-negative bacteria (Sasser, 1990). Expert systems consisting of capillary gas chromatograph, integrator, computer and component analysis and pattern recognition software are now available which allow the automatic identification of a large number of different bacterial species (Sasser, 1990) by their fatty acid profiles.

Polar lipids which can be readily extracted as free lipids from the cells produce 'fingerprint patterns' on TLC which are characteristic.

Also the respiratory quinones localized in the plasma membranes, the menaquinones and ubiquinones, produce species and group specific patterns (Collins, 1985).

Analysis of Nucleic Acids

The guanine + cytosine content of DNA. Only a few years after the elucidation of the structure of DNA by Watson and Crick in 1953, it was discovered that the base composition of bacterial deoxyribonucleic acids varies widely among different bacterial species. The molar percentage of the DNA bases guanine plus cytosine was called the GC value. It is one of the very few constants in biology, of great important for classification and identification.

The GC range among bacteria ranges from approx. 25 mol% to approx. 75 mol% (Austin and Priest, 1986). This very broad span of GC values is characteristic for prokaryotes. It reveals the great evolutionary diversity of this biological group. Other more highly evolved groups of eukaryotes like the vertebrates, for example, have a quite narrow GC range.

If there is a substantial difference in GC values between two organisms they cannot possibly be closely related. Within a species GC values should not differ by more than 1 to 3 mol% (Kurtzman, 1985) and differences within a genus are usually less than 10 mol%. Two strains which differ by 10% or more hardly have any DNA consequences in common and therefore cannot be phylogenetically related. The reverse, however, is not necessarily true: two strains may have identical GC contents and yet are not closely related. The GC content has its value to exclude close relationship.

There are many methods available to measure the GC contents, but only a few of them are practicable for routine work.The classical methods for the determination of the GC contents are equilibrium density centrifugation, chemical degradation and subsequent paper chromatography and spectroscopical methods (Mandel and Marmur, 1968;Johnson, 1981).

Spectroscopical methods are still most widely used but reversed-phase high-performance liquid chromatography is becoming more and more popular as a fast and reliable technique for the G+C determination of bacterial DNA (Meshbah *et al.*, 1989). DNA is hydrolyzed with P1 nuclease, dephosphorylized with bovine alkaline phosphatase. The resulting nucleosides are separated by chromatography on a reversed phase analytical column. The analysis requires very little DNA, is rapid and allows the detection of unusual nuclear bases e.g., methy1-deoxycytidine, hydroxy-methyl-uridine and methy1-adenosin.

Indirect determination of nucleic acid sequence homology by hybridization. Nucleic acid hybridization techniques depend on the properties of single stranded DNA (or RNA) molecules to recognize other complementary single stranded molecules and form duplexes with them *in vitro.* In other words, hybridization allows the detection of long stretches of genetic homology between DNA or RNA molecules of different organisms. In principle there are two types of nucleic acid hybridization: in the 'free solution' systems the molecules both react in solution and in the 'mixed phase' systems one of the partners is fixed on a solid support, whereas the other partner is in solution. All methods, except the spectroscopic methods, require that one of the two DNA reactants is labelled, either with radioactive isotopes (^{3}H, ^{125}I, ^{14}C, ^{32}P) or chemically (e.g., biotin-11-dUTP, photobiotin, digoxigenin). The hybrid molecules formed are detected either by radiometry, spectroscopy, colorimetry, or by fluorescence, luminescence techniques.

Spectroscopic DNA-DNA hybridizations require a special spectrophotometer allowing simultaneous control of optical density, temperature, and time: these parameters must be recorded simultaneously for normally 4 cuvettes.

The spectroscopic DNA-DNA hybridization technique of De Ley *et al.* (1970) is well established and widely used in bacteriology. It is a relatively simple, fast technique requiring only the recording of the initial phase of the DNA renaturation. The method's most prominent drawbacks are its low resolution power at low homology values and relatively 'high background' (Huss *et al.*, 1983; Jahnke and Bahnweg, 1986) resulting in homology values of 20-40%, even if little DNA sequence homology exists.

Nucleic acid hybridization on membrane filters is probably the most often used hybridization procedure in molecular biology. Here one of the two single-stranded nucleic acids is fixed on a membrane filter, whereas the other is in solution.

The interpretation of DNA-DNA homology data obtained from DNA-DNA hybridization experiments is not always easy. These data depend very much on the method and the specific experimental conditions.

A direct comparison of data obtained with different methods is not always possible.

Bacteriologists tend to place the species boundaries at rather low DNA-DNA homologies of 65-70%. In yeast taxonomy values of 80 to 100% are believed to indicate conspecificity.

Because of the low resolution power of the fast spectroscopic method at low DNA-DNA homologies, no detailed interpretation of values below approx. 50% is recommended. Low values just mean 'not closely related'.

DNA restriction analysis. Another approach to physical characterization of DNA is restriction analysis and Southern blotting of DNA. By these techniques 'DNA fingerprints' are generated which may serve for the identification of closely related organisms. With the development of pulsed fields gel electrophoreses techniques also whole bacterial chromosomes or very large chromosomal fragments can be analyzed.

RNA sequencing. The analysis of the ribosomal DNA offers resolution above species level, because the ribosomal RNA genes have been highly conserved during evolution. There are three types of ribosomal RNA which are distinguished by size. The 5S rRNA is 116-120 nucleotides long, whereas 16S rRNA and 23S rRNA comprise about 1600 and 3300 base pairs, respectively. rRNA analysis can be carried out by hybridization of isolated 16S or 23s rRNA to chromosomal DNA, or by partial or complete sequencing.

Because of the small size many bacterial 5S rRNAs have been sequenced (Pace *et al.*, 1985). In these small molecules, however, mutational base changes have a strong influence on the results of dendrograms calculated from the sequences.

For this reason the larger 16S molecules were finally chosen for comparative cataloguing (Stackebrandt and Woese, 1984). In the classical procedure, 16S rRNA is cleaved into oligonucleotides, labelled radioactively and then separated by two-dimensional paper and compared to the sequences of other organisms and a similarity coefficient is calculated.

With new sequencing techniques available, it is now possible to sequence large stretches of rRNA sequences directly (Pace *et al.*, 1985) and thus obtain intact and complete 16S rRNA sequences.

16S rRNA sequencing has provided profound insight into the natural (phylogenetic) relationships of the bacterial world. It is clear that a deep genetic split exists among the bacteria, dividing them into the Archaeobacteria and the True Bacteria, the Eubacteria (Fox et al., 1980). These two lines are as different from one another as eukaryotes are from both lines.

Although superficially similar, the Archaeobacteria are quite different from the Eubacteria by their cell walls, their membrane composition, their transfer RNA composition and other important chemotaxonomic markers. Archaeobacteria are, for example, methanogenic, halophile and thermoacidophile bacteria, whereas the True Bacteria are those organisms that have been the normal objects of microbiology.

CONCLUSIONS

For proper bacterial identification, classical microbiological tests and modern chemical techniques must both work together. The classical tests and morphological observations provide the basis for the determination of the bacterial groups and in many cases will also allow identification.

If identification fails, more sophisticated chemical analytical methods are used which, in many cases, may allow the final identification of the unknown organism. It is important, however, to remember that only classified organisms can be identified. If no unequivocal (clearcut) identification result can be reached, it is possible that the unknown organisms belong to a new as yet undescribed species.

REFERENCES

Austin, B. and Priest, F. 1986. *Modern Bacterial Taxonomy.* Van Nostrand Reinhold Co., Wokingham, England.

Claud, D. 1981. *Characterization and Identification of Bacteria.* Manual of the Training Course on Running and Management of Culture Collections. Brno, Czechoslovakia.

Collins, M.D. 1985. Isoprenoid quinone analysis in bacterial classification and identification. In M. Goodfellow and D.E. Minnikin (eds), *Chemical Methods in Bacterial Systematics.* Academic Press, London. pp. 267-88.

Cowan, S.T. and Liston, J. 1974. The mechanism of identification. In R.E. Buchanan and N.E. Gibbons (eds), *Bergey's Manual of Determinative Bacteriology.* The Williams and Wilkins Company, Baltimore, Maryland USA. 8th end pp. 10-13.

De Ley, J., Cattoir, H. and Reynaerts, A. 1970. The quantitative measurement of DNA hybridisation from renaturation rates. *Eur. J. Biochem.*, 12, 133-42.

Fox, G.E., Stackebrandt, E., Hespell, R.B., Gibson, J. Maniloff, J., Dyer, A., Wolfe, R. S., Balch, W.E., Tanner, R.S., Magrum, L.J., Zablen, L.B., Blakemore, R., Gupta, R., Bonen, L., Lewis, B.J., Stahl, D.A., Luehrsen, K.R., Chen, K.N. and Woese, C.R. 1980. The phylogeny of prokaryotes. *Science*, 209, 457-63.

Gutteridge, C.S., Vallis, L. and MacFie, M.J.H. 1985. Numerical methods in the classification of micro-organisms by pyrolysis mass spectrometry. In M. Goodfellow, D. Jones and F.G. Priest (eds), *Computer-assisted Bacterial Systematics.* Academic Press, London. pp. 389-401.

Helm, D., Labischinski, H., Schallehn, G. and Naumann, D. 1991. Classification and identification of bacteria by FT infrared spectroscopy. *J. Gen. Microbiol.*, 137, 69-79.

Huss, V.A.R., Festl, H. and Schleifer, K.H. 1983. Studies on the spectroscopic determination of DNA hybridization from renaturation rates. *System. Appl. Microbiol.*, 4, 184-92.

Jackman, P.J.H. 1985. Bacterial taxonomy based on electrophoretic whole cell protein patterns. In M. Goodfellow and D.E. Minnikin (eds), *Chemical Methods in Bacterial Systematics.* Academic Press, London. pp. 115-30.

Jahnke, K. D. and Bahnweg, G. 1986. Assessing national relationships in the Basidiomycetes by DNA analysis. *Tr. Br. Mycol, Soc.*, 87, 175-91.

Johnson, J.L. 1981. Genetic Characterization. In N.R. Krieg (ed), *Manual of Methods for General Bacteriology*, 22. American Society for Microbiology, Washington, DC, USA. pp. 450-72.

Kurtzman, C.P. 1985. Molecular taxonomy of the fungi. In J.W. Bennett and L.L. Lasure (eds), *Gene Manipulations in Fungi.* Academic Press, New York, USA. pp. 35-56.

Krieg, N.R. and Holt, J.G. (eds). 1984. *Bergey's Manual of Systematic Bacteriology.* Williams and Wilkins, Baltimore, London. Vol. 1.

Lapage, S.P. Sneath, P.H.A., Lessel, E.F., Skerman, V.B.D., Seelinger, H.P.R. and Clark, W.A. 1975. Int. Code of Nomenclature of Bacteria and Statutes of the Int. Committee on Systematic Bacteriology and Status of the Bacteriology Section of the Microbiol. Sciences. Am. Soc. for Microbiology, Washington, DC, USA.

Mandel, M. and Marmur, J. 1968. Use of ultraviolet absorbance temperature profile for determining the guanine plus cytosine content of DNA. In L. Grossman and K. Moldave (eds), *Methods of Enzymology.* Academic Press, New York, USA. Vol. XII B, pp. 195-206.

Meshbah, M., Premachandran, U. and Whitman, W. B. 1989. Precise measurement of the G+C content of deoxyribonucleic acid by high-performance liquid chromatography. *Int. J. Syst. Bact.*, 39, 159–67.

Naumann, D., Helm, D. and Labischinski, H. 1991. Microbial characterizations by FT-IR spectroscopy. Nature, 351, 81-82.

Pace, N.R., Stahl, D.A., Lane, D.J. and Olsen, G.J. 1985. Analyzing natural microbial populations by rRNA sequences. *American Society for Microbiology News*, 51, 4-12.

Sasser. 1990. Identification of bacteria by gas chromatography of cellular fatty acids. *US Fed. Culture Collections Newsletter*, 20, 1-6.

Schleifer, K.H. and Kandler, O. 1972. Peptidoglycan types of bacterial cell walls and their taxonomic implications. *Bacter. Reviews*, 36, 407-77.

Sneath, P.H.A., Mair, N.S., Sharpe, M.E. and Holt, J.G. (eds). 1986. *Bergey's Manual of Systematic Bacteriology*. Williams and Wilkins, Baltimore, London. Vol. 2.

Stackebrandt and Woese. 1984. The phylogeny of prokaryotes. *Microbiol. Sciences*, 1, 117-22.

Staley, J.T., Bryant, M.P., Pfenning, N. and Holt, J.G. (eds). 1989. *Berger's Manual of Systematic Bacteriology*. Williams and Wilkins, Baltimore, London. Vol. 3.

Williams, S.T., M.E. and Holt, J.G. (eds). 1989. *Bergey's Manual of Systematic Bacteriology*. Williams and Wilkins, Baltimore, London. Vol. 4.

13

MICROBIAL STRAIN DATA MANAGEMENT AND BACTERIAL IDENTIFICATION USING IBM-COMPATIBLE MICROCOMPUTERS

K.D. JAHNKE

DSM—German Collection of Micro- organisms and Cell Cultures, Braunschweig, Germany

Summary. In microbiology, many data concerning the description and the performance of micro organisms are collected and need to be stored properly so that they can be quickly retrieved and analyzed. The advances in computer technology, computer software and most notably the decrease in costs have made the personal microcomputer the universal tool for every scientist. There are numerous applications for computers in microbiology, not only for text processing, but mainly for collection of data in databases and for comparing and analyzing these data in microbial identification and characterization schemes.

The Microbial Information System MICRO-IS is presented as an example of a non-relational integrated shareware database and identification system. It was developed for microbiologists in the USA and has been tailored to facilitate the management of microbiological data. MICRO-IS is useful for any laboratory involved in microbial identification or microbial strain data collection.

MICRO-IS displays all the major features of commercial database management systems and can handle very large data sets of several megabyte. In contrast to other commercial systems, however, MICRO-IS includes an identification module which will allow probabilistic identification of bacteria using one of the 22 included data matrices of bacteria or any other user-defined data matrix.

INTRODUCTION

The advances in the world of personal microcomputers have made it possible to obtain personal computers for the microbial laboratory at comparatively low prices. More and more computer programmes become available dealing with data storage and microbial identification.

Some programmes are distributed as shareware for scientific research (McMagnus *et al.*,1990; Mendt and Georgi, 1990; Flandrosis *et al.*, 1991; Kellogg, 1979) or commercially (cf list of low cost or shareware systems; Pitt, 1991; Barnett et al., 1990; Bryant, 1991) at modest prices. Besides these programmes which are more oriented towards basic research and low budget institutes, large and expensive commercial expert systems exist for the use in industrial companies and hospitals (cf list of commercial expert systems).

In this paper, the properties of some of the shareware and low cost programmes will be discussed. No attempt has been made to cover the subject completely. There is no doubt that in this quickly growing field of computer software, some interesting programmes have been overlooked. An excellent and comprehensive list of programmes has been compiled by Sackin (1987) and is recommended as a help to potential users who want to find out which programme to use for classification and identification of micro-organisms.

WHAT IS A DATABASE ?

A computer database is a collection of information stored in one or several computer files. The database may contain different types of information such as literature data or strain information.

One definition of a database given by Painting (1989) reads like this:

> A database is a well organised collection of data contained in one or more files. One should be able to process, update and

make additions to the contents of the database in a simple and flexible way. It should be easy to make different kinds of unplanned as well as planned retrievals of data from the database.

There are six essential operations involved in working with computer databases:

1. data addition
2. data checking
3. data modification
4. database restructuring
5. data searching
6. creating reports/extracts from the database.

The Structure of a Database

What structure a database should have depends very much on the needs of the user and the type of data to be stored. Let us take a telephone book as an example for a database.

One data record in the book consists of the name of the person, the street he lives in and his telephone number. The information (letters and numbers) is stored in *fields*, here called NAME, STREET and TEL NO. The names of the fields are unique and must not be used for other fields. All records are combined into a *file* called CITY.TEL, in our example the file is the city telephone book. In some databases like MICRO-IS it is possible to divide the database into sub-bases called tables. In our example the records of people living in suburbs may be stored in tables called SUB1, SUB2 AND SUB3.

The information in the telephone book is well organized and indexed by the name of a person. Retrieval of information, however, is possible only by the alphabetical order of names, Not so in a computer database. Here the data may be searched by any kind of field (e.g., STREET) or the data may be indexed by TEL NO instead of NAME. Extracts can be made. For example: list all data records (= NAME+STREET+TEL NO)of people called 'Miller' and save the data in a separate file called MILLER.TEL.

Types of Databases

The simplest form of a database is a text file created by a computer text editor. Here the data are stored in text tables and the 'record' is a line in the table, a 'field' is a column. Information is retrieved with the SEARCH function of a word processor or by using the DOS command FIND. This textfile 'database', however, has severe limitations. Searching is comparatively slow and data manipulations are severely restricted. Most professional databases use special database management software which has been adapted to the special needs of the user. Examples for commercial database software are DBASE (Ashton-Tate Corp., USA), PARADOX (Borland Inc., USA) and DATAPERFECT (Word Perfect Corp., USA). All these are powerful programme packages but they need to be 'programmed' for the special needs of microbiology. Interactive user surfaces have to be created, data entry forms designed, report forms have to be made, etc.

These programmes are relational databases. The word 'relational' refers to the special way the information is stored in the databases. As the term suggests, filed holding data re related to one another. The different files are physically separate but are linked by one or more common fields. In a microbiological strain database this common field could be the strain number. For example, the data file MORPH.DAT holds morphological data, PHYSIOL.DAT holds physiological data, and GENET.DAT holds the genetical data of bacterial strains. The files are linked by the common field STRAIN. Relational databases are more compact compared to non-relational databanks such as MICRO-IS and offer more possibilities of data manipulation.

If you are designing a database, you have to decide on the number of fields in a record, the type of the field (e.g., character field, numeric field, binary field, date field), the name of the field and the size of the field (number of digits). Considerable time should be spent at this early stage because hasty decisions may cause problems later on in the project.

The Database System MICRO-IS 5.0

The Microbial Information System MICRO IS is a non-relational integrated shareware database and identification system which was developed in the USA for microbiologists. MICRO-IS displays all the major features of commercial database management. It was created originally as a version for main frame computers, was adapted to MS-DOS compatible microcomputers and was made available to the general scientific community as a shareware system. It does not need to be 'programmed' but is a ready to use system.

Simple Strain Catalogue Databank Using MICRO-IS 5.0

It is beyond the scope of this paper to give a detailed introduction to MICRO-IS. Instead, the creation of a simple strain data bank is described. For more information on the MICRO-IS system, the reader is referred to the MICRO-IS manual distributed on computer disk along with the programme by Microbial Strain Data Network (MSDN) in Cambridge, England (cf list of low cost of shareware systems).

The databank CATALOG containing the table TAB1 with the following fields is created using the DEFINE/CREATE command:

STRAIN is the key field and contains the unique strain number of the organism in the collection (e.g., 2366). The key field must be a 'character field' and cannot be any other type of field. GENUS (e.g., Cytophaga) and SPECIES (e.g., heparina) contain the scientific name of the authorities who created the species name and the year of the publication (e.g., Payza and Korn, Christensen, 1980). SYN-COLL contains the collection numbers of other culture collections (e.g., AT 13125, IFO 12017, NCIB 9290). HISTORY holds the names of persons (or institutions) from whom the culture was obtained and the original strain designation (e.g., W. Mannheim, HIM762-3). LIT holds the numbers of the literature references (e.g., 1778, 1179).

TABLE: 13.1
FIELDS OF THE MICRO-IS DATABASE CATALOG

Field number	*Field name*	*Field length*	*Field type*	*Key field*
C00001	STRAIN	10	Character	Yes
C00002	GENUS	20	Character	No
C00003	SPECIES	20	Character	No
C00004	AUTHORS	20	Character	No
C00005	SYN-COLL	80	Character	No
C00006	HISTORY	80	Character	No
C00007	LIT	80	Character	No
C00008	REMARKS	80	Character	No

Finally, in the field REMARKS there is room for miscellaneous information not fitting into one of the other fields: fields such as data fields on organisms properties (see below) may be added at a later date if necessary.

This 'electronic catalogue' may be searched with the QUERY utility using the LIST command. Some examples of the use of LIST are given in the section below on synoptic keys. Simple reports may be created using the SET DESTINATION and the LIST commands. The resulting text files, however, have to be edited by a text processor before a decent catalogue can be printed.

THEORY OF IDENTIFICATION

Identification or diagnosis is a comparative process by which an unknown organism is compared to known organisms. It may well happen, as is often the case, that an unknown organism is isolated which cannot be identified with the presently known databases. Even the best and most powerful programmes cannot identify an undescribed organism. So some caution and foresight should be exerted when using computer programmes. Before consulting the computer, the user should have checked thoroughly if the right microbial group has been determined and the correct database has been chosen.

DICHOTOMOUS KEYS

These types of keys are still widely used in botany, zoology, and mycology but rarely in microbiology. In dichotomous keys, the presence or absence of a character is checked sequentially. Each decision ('character present' or 'character absent') leads to a new part of the key, so moving the user closer to the final goal of identification (Fig. 13.1). The inherent drawback of the system is that one wrong decision may make identification impossible.

TABLE:13.2
EXAMPLE OF A SYNOPTIC IDENTIFICATION TABLE

Species	*CHAR_A*	*CHAR_B*	*CHAR_C*	*CHAR_D*
Species 1	+	+	-	-
Species 2	-	+	+	-

Synoptic Keys

Synoptic character tables can be entered into normal databanks and may be used for identification by searching for exact or approximate matches. DBASE (Ashton-Tate Corp., USA) and the database system ASK SAM ('Access Stored Knowledge via Symbolic

Access ' Method', North American Software, USA) have been used successfully for this purpose (Pitt, 1991; Kendrick, 1990); however, unless the system has been programmed with a good 'user surface' searches and identifications are comparatively complex and cumbersome.

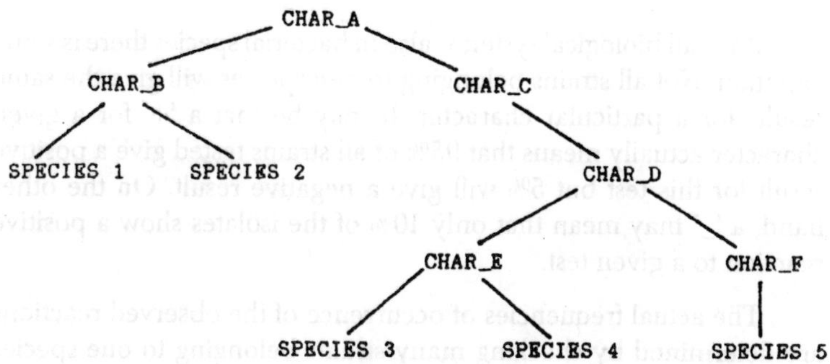

Fig. 13.1: Example of a dichotomous identification key

The known species and their characters are grouped in a table as shown in Table 13.2. The presence or the absence of a character is marked by a '+' or '-', or if the result is unknown the test is marked with '?'. Now the user writes down the results of the unknown on a separate piece of paper and scans through the whole character table looking for the species to match the results of the unknown. If the search is not successful, at least the species which comes closest to the unknown can be found. Working through a large table by this method requires some practice and is prone to errors.

Synoptic searches can be performed by computer databank programmes like MICRO-IS. Here the databank utility QUERY is used to perform the identification. The QUERY comment LIST and the operations AND, OR, NOT, ANY offer many possibilities for identification searches. The fields CHAR_A to CHAR_D are so-called binary fields which contain either '1' (= positive) or '0' (= negative) or a 'space' for unknown results. For example:

- Find exact match (all characters positive)
  LIST (GENUS SPECIES)
  (CHAR_A AND CHAR_B AND CHAR_C AND CHAR_D))
- Find approximate match, two tests different, either negative or unknown

LIST (GENUS SPECIES)
(ANY (2 CHAR-A CHAR-B CHAR-C CHAR-D))

If the QUERY command lines are very long and complex, it is advisable to store them in separate text files which can be edited and used again by the QUERY utility.

Probabilistic Identification

As in all biological systems, also in bacterial species there is some variation. Not all strains belonging to one species will give the same results for a particular character. It may be that a '+' for a given character actually means that 95% of all strains tested give a positive result for this test but 5% will give a negative result. On the other hand, a '_' may mean that only 10% of the isolates show a positive reaction to a given test.

The actual frequencies of occurrence of the observed reactions are determined by checking many strains belonging to one species and arranging the results in probability tables as shown in Table 13.3. In a way probability tables are condensed extracts of data banks reflecting the original (binary) data obtained with the separate strains. The author of the matrix has to decide which of the strains actually belong to one species and which do not, he then calculates the frequencies of positive results of the different characters.

TABLE: 13.3
EXAMPLE OF A PROBABILISTIC IDENTIFICATION TABLE

Likelihood of positive result for: *Species*	*CHAR_A*	*CHAR_B*	*CHAR_C*	*CHAR_D*
Species 1	0.80	0.99	0.05	0.01
Species 2	0.10	0.99	0.95	0.01

Lapage *et al.* (1970, 1973) developed a computer-based identification programme based on Baye's Theorem which calculated the likelihood (probability) of an unknown bacterium belonging to one of the species included in the identification matrix. An important prerequisite for the use of this system is the fact that the tests must be independent of one another. In reality this may not be the case because some characters may depend on enzymes of a common metabolic pathway. These characters are called linked and may seriously influence identification results. If a new test set for identification is developed, the usefulness of the tests may be tested and the minimum number of diagnostic characters may be selected by the computer programmes CHARSEP (Sneath, 1979) and DIACHAR (Sneath, 1980a).

The probability of each character being positive is obtained by multiplying the probabilities of the individual and independent test results by one another (Table 13.4). These 'row probabilities' are then normalized, that is, they are divided by the sum of all row probabilities. The resulting number is called the 'normalized identification score'. It is an indicator on how closely an unknown strain resembles the species in the matrix. An identification score of 0.99 indicates that the unknown resembles the species in question.

Since the 'normalized ID-Score' is a relative value it may lead to false identification if an insufficient number of tests has been chosen, or the wrong matrix is used or, simply, because the chosen matrix does not include the unknown organism. Also, best results are only obtained if a *complete* test set with the maximum number discriminative tests is used. A good identification would not only result in a high ID Score of 0.99 (or 99%), but also the actual row probability of the unknown strain would be very similar to the row probability of a *typical* strain of that species (Table 13.4).

A *typical* strain may be defined as the 'Hypothetical Median Organism' (HMO) prepared from positive characters occurring in 50% or more of the samples (Tsukamura nd Mizuno, 1968). The best ID scores of a typical strain of each bacterial group may be calculated by the computer programme MOSTTYP (Sneath, 1980b).

MICRO-IS also calculates the so-called 'best likelihood' which is the row probability of a *typical* strain of each species calculated by the procedure shown in Table 13.4. MICRO-IS uses the ratio 'Best Likelihood of Typical Strain/Likelihood of Unknown' as an additional indicator for the quality of an identification. A strain is considered to be identified only if the ID score is greater than or equal to 0.99 (or 99%) *and* the ratio 'Best Likelihood/Likelihood' is equal to or smaller than 100.

If the unknown strain has several characters in common with other strains in the matrix the ID-score will drop. If the ratio 'Best Likelihood/Likelihood' is near 1 and the ID-score about 0.8 (or 80%) the identification may still be correct.

The statistical approach has been frequently used in m lical and microbiological diagnosis (Dybowski and Franklin, 1968; W lcox *et al.*, 1973; Wilcox and Lapage, 1975; Friedman and Maclowry, 1973; Friedman *et al.*, 1975). Since this method can be readily applied to most groups of bacteria and yeasts, it is probably the most widely used system and has a considerable potential for commercialization.

TABLE 13.4

CALCULATION OF ROW PROBABILITIES, IDENTIFICATION SCORES, AND BEST LIKELIHOODS.

Likelihood of positive result for Species	*CHAR_A*	*CHAR_B*	*CHAR_C*	*CHAR_D*
Species 1	0.80	0.99	0.05	0.01
Species 2	0.10	0.99	0.95	0.01
Unknown isolate	+	+	-	-
Row probability Species1	0.8 × =0.7449	0.99 ×	(1-0.05) ×	(1-0.01)
Row probability Species 2	0.1 × 0.99 × (1-0.95) × (1-0.01) = 0.00490			
Sum	0.7449 + 0.00490 = 0.7498 = 100%			
Normalized ID score (%) for the unknown to be identical to:				
Species 1	99.3			
Species 2	0.7			
Typical Species 1	+	+		-
Best likelihood	0.8 × 0.99 × (1-0.05) × (1-0.01) = 0.7449			
Typical Species 2	-	+	+	-
Best likelihood	(1-0.1) × 0.99 × 0.95 × (1-0.01) = 0.83799			
Comparison of unknown to typical Species 1	Likelihood of unknown/ best likelihood		0.7449/0.7449 = 1.00	
Comparison of unknown to typical Species 2	Likelihood of unknown/ best likelihood		0.83799/0.7449=1.12	

The probabilistic approach to identification can be programmed with relative ease in various high-level computer programming languages like PASCAL or even BASIC. The inherent drawback of these programmmes, however, is the fact that they need estimates of probabilities for each character of a species. These probability tables are only available for a limited number of bacteria and yeasts but usually not for groups of filamentous fungi or plants.

Fungi, for example, can be better described by synopitc character tables (Tables 13.2). There have been attempts for yeasts to transform synopitc character tables into probability tables by substituting a '+' for 90%, and '-' for 10% and a 'variable' character into 50% probability. These 'pseudo-likelihood' matrices are useless and will not give reliable identifications when fed into probabilistic identification programmes.

Other Identification Schemes

An alternative for filamentous fungi to the above-mentioned systems may be the 'maximum detection system' (Mendt and Georgi, 1990) which has been tested successfully on the gilled fungi, the Agaricales. Species characters are coded and given points corresponding to their (estimated) frequencies of occurrence. For example, the diameter of the fruiting body of a given species may lie in the range 4-12 cm. Fruiting bodies having 7-9 cm occur most often and are given 3 points. Fruiting bodies having 5-7 or 10-12 cm are less common and are given 2 points, those between 4 and 5 cm and 11 and 12 cm in diameter are rare and given only 1 point. For each fungal species a list of coded characters exists, which are collected in a database and each time the characters agree, the corresponding points are added up. Species similar to the unknown will have a high number of points, dissimilar species a lower number.

If an identification system is to be developed from rough and imprecise date from publications, the programme EXPRIM (Flandrois *et al.*, 1991) may be employed. EXPRIM is an 'inference engine', a special kind of database devoted exclusively to identification. One major advantage of the system is its independence from numerical data. Although the use of the system looks somehow complicated, it seems to be applicable to bacterial and fungal identification when exact probability tables ae missing.

CONCLUSIONS

Computer database systems and identification programmes offer appreciable help in various fields of microbiology and mycology. They are especially helpful when large numbers of observations are to be compared for the purpose of identification. Computer programmes, however, cannot substitute for exact observations and skills of the experimenter who must produce the date the computer needs in order to make a correct diagnosis.

REFERENCES

Dybowski, W. and Franklin, D.A. 1968. Conditional probability and the identification of bacteria: a pilot study. *J. Gen. Microbiol.*, 54, 215-29.

Flandrois, J.P., Carrett and Pave, A. 1991. EXPRIM: An expert system useful in the phenotypic identification of bacteria. *BINARY*, 3, 53-56.

Friedman, J.P. and Maclowry, J. 1973. Computer identification of bacteria on the basis of their antibiotic susceptibility patterns. *Appl. Microbiol.*, 26, 314-17.

Friedman, R.B., Bruce, D., Maclowry, J. and Brenner, V. 1975. Computer-assisted identification of bacteria. Am. *J. Clin. Pathol.*, 60, 395-403.

Kellogg, S.T. 1979. MICRID: a computer-assisted microbial identification system. *Appl. Environm. Microbiol.*, 38, 559- 63.

Kendrick, W.B. 1990. A rapid computer-based synopitc key to genera of gilled fungi. *Mycotaxon*, 38, 27-36.

Lapage, S.P., Bascomb, S., Willcox, W.R. and Curtis, M.A. 1970. Computer identification of bacteria. In A. Baille and R.J. Gilbert (eds), *Automatization, Mechanization and Data Handling in Microbiology*. Academic Press, London. pp. 1-22.

Lapage, S.P., Bascomb, S., Willcox, W.R. and Curtis, M.A. 1973. Identification of bacteria by computer: general aspects and perspectives. *J. Gen. Microbiol.*, 77, 273-90.

McMagnus, C., Blaine, L. and Krichevsky. 1990. *Self Instruction Manual for the Micro-IS, Version 4.11*. Microbial Systematics Section, National Institute of Dental Research, Bethesda, USA.

Mendt, A. and Georgi, M. 1990. Pilzbestimmung mit Hilfe der Maximum-Detektion. *Z. Mykologie*, 56, 187-98.

Painting, K. 1989. Introduction to Databases. UKFCC Data Management for Culture Collections Course, 4-6 July 1989, Norwich, England.

Pitt, J.I. 1991. Using dBase to write PENNAME, a computer key to common penicillium species. *J. Appl. Bact.*, 71, 86-91.

Sackin, M.J. 1987. Computer programs for classification and identification. In R.R. Colwell and R. Grigorova (eds), *Current Methods for Classification and Identification of Micro organisms*. Academic Press, London.

Sasser, 1990. Identification of bacteria by gas chromatography of cellular fatty acids. *US Federation for Culture Collections Newsletter*, 20, 1-6.

Sneath, P.H.A. 1979. BASIC program for character separation indices from an identification matrix of percent positive characters. *Computers & Geosciences*, 5, 349-57.

Sneath, P.H.A. 1980a. BASIC program for the most diagnostic properties of groups from an identification matrix of percent positive characters. *Computer's & Geosciences*, 6, 21-26.

Sneath, P. H. A . 1980b. BASIC program for determining the best identification scores possible from the most typical examples when compared with an identification matrix of percent positive characters. *Computers & Geosciences*, 6, 27-34.

Tsukamura, M. and Mizuno, S. 1968. Hypothetical mean 'organism' of mycobacteria. *Jpn J. Microbiol.*, 12, 371-84.

Willcox, W.R. and Lapage, S.P. 1975. Methods used in a program for computer-aided identification of bacteria. In R.J. Pankhurst (ed). Academic Press, London. pp. 103-19.

14

COLLECTION AND PRESERVATION OF MICRO-ORGANISMS: SOME NEW DEVELOPMENTS

K.A. MALIK

DSM—German Collection of Micro-organisms and Cell Cultures, Braunschweig, Germany

Summary. Biotechnologically significant micro-organisms are being applied increasingly to solve agricultural, food, health, energy, and environmental problems, and there is an ever growing need for the collection, preservation, and storage of newly isolated strains, genetically engineered strains and strains with plasmids.

Preservation is mainly performed to keep cultures viable, uncontaminated, and unchanged in their properties (biochemical, morphological, physiological, genetic, etc.). Based on these criteria, various simple, modern, and miniaturized methods are available for the preservation of micro-organisms.

Several major laboratories and culture collections with specialist back-up use sophisticated and modern methodology of preservation.

In addition, research is conducted in order to develop new methods and optimize the preservation protocols with the aim of increasing their applicability to a wide range of organisms and overcome problems like longterm viability, stability, improved shelf-life, storage space, and sensitivity of new strains to existing preservation processes.

Deep-freezing at-198°C in liquid nitrogen, freeze-drying (lyophilization) and liquid drying (L-drying) are the modern methods that to various degrees secure viability and strain stability during preservation.

For the long-term preservation of micro-organisms in small laboratories where such arrangements are not available, some simple successful methods of drying and freezing are recommended for a wide variety of micro-organisms. This article describes some recent developments in such preservation methods.

INTRODUCTION

The maintenance of micro-organisms in a viable and stable state is an essential prerequisite for microbiology and biotechnology. Cultures are collected and maintained for a variety of reasons by teaching and research laboratories; by industrial laboratories; medical laboratories; and by genebanks.

However, despite this fundamental importance of a reliable supply of pure and stable cultures, culture preservation is often afforded low priority and is commonly carried out with inadequate equipment and staffing. Similarly, where culture maintenance is not a primary function, barely any time is devoted to develop or assess different preservation methods.

This vital function is mainly carried out by service culture collections that have a full commitment to the preservation of cultures and their accumulated expertise is invaluable to others.

In recent years, successful culture preservation has reached a high level of success and has been a major achievement in biology (Kirsop and Doyle, 1991).

Although a number of excellent preservation techniques have now been developed, they must be correctly applied if contamination, loss of viability, and genetic change are to be prevented.

Thus the selection of maintenance methods that produce good survival levels and strain stability is of fundamental importance. Practical details on a few selected simple and sophisticated methods for the long-term preservation of micro-organisms is described in the course manual on *Culture Collection and Strain Management*, Prague,

1991. The methods work well and are reliable with regards to expected survival levels, shelf-life and stability of preserved micro-organisms. These methods may also prove useful for other cell types than for a particular group of tested micro-organisms for which these have been found successful.

The methods employed at the DSM are mostly used for biotechnologically important collection and so provide a suitable model for long-term storage of many sensitive and delicate micro-organisms (Malik, 1984, 1985, 1987, 1988, 1990, 1991).

For short-term storage, almost all micro-organisms can be maintained as living cultures on appropriate media and need to be transferred after a definite interval as depending upon the nature of the strain and the growth conditions.

However, for sensitive microbes, continuous attention is needed during such subculturing, and a bulk storage of replicates demands space and labour. Maintenance of cultures through subculturing is least satisfactory because of genetic drift and various other problems.

Preservation by drying, freezing, and freeze drying offers some stability and facilitates long-term storage. Freezing is used for those micro-organisms that do not survive freeze-drying and is a preferred method for long-term storage of micro-organisms at culture collections.

Through dehydration, metabolic rates of organisms can be reduced but simple drying is not recommended for long-term storage of all micro-organisms. Drying from the liquid state avoiding freezing, is called L-drying and is a successful and effective technique for the long-term preservation of such cultures which are sensitive to freezing or freeze-drying. This method has several advantages over freeze-drying and has has been successfully applied in the preservation of various sensitive micro-organisms.

Cryopreservation and freeze-drying are widely used techniques and there are many variations in technical details.

Information on some newly developed methods will be provided in this article which could be used as a suitable model for the preservation of a wide range of micro-organisms.

LIQUID-DRYING OF MICRO-ORGANISMS

Liquid-drying is generally known as L-drying (Annear, 1962). It involves vacuum drying of samples from the liquid state without freezing. It is a succcessful method for many delicate micro-organisms especially for such cultures which are sensitive to freezing or freeze-drying (Banno and Sakane, 1979; Malik, 1990a). Although this method has several advantages over freezing and freeze-drying, it has seldom been used in small laboratories as in the conventional method specialized equipment and continued care is required to create the conditions conducive to the liquid-drying process.

Simplified Liquid-Drying Process

Conventionally, the cell suspensions to be dried are usually added on to various carriers in vials which are then attached to manifold nipples and lowered into water bath at 20°C. During L-drying, the drying process is accelerated by applying a vacuum and it is so manipulated that no freezing should occur. For good survival and stability, protective agents are used and a dense suspension of the culture is prepared in such suspending fluids. In principle, freezing can be avoided by various means, one being to use small volumes of cell suspension spread over large surface areas. Such surface areas which are good heat conductors prove helpful in avoiding freezing.

This method has recently been improved and simplified by the author (Malik, 1990a; 1990c). In this method (Fig. 14.1) activated charcoal has been used along with various protective agents such as skimmed milk and meso-inositol, honey, glutamate, or raffinose (Malik, 1976, 1988a).

In contrast to the conventionally used carriers materials (which are often bad thermal conductors), activated charcoal proved to be an excellent thermal conductor and it ensured proper distribution of heat in the samples, so that throughout the procedure the temperature of the cell material could be maintained at 20°C. In combination with other additives used, activated charcoal served as an ideal absorption base and almost no frothing occurred, even during drastic evacuation and degassing.

At the DSM, this method has successfully been used for the preservation of various sensitive aerobic and anaerobic micro-organism. This method has also proved effective for the successful preservation of freshwater unicellular green algae (Chlorococcales) and thus is first of its kind even applicable for the drying of eukaryotic algae (Malik, 1991e). The apparatus used at the DSM for the new liquid-drying process is shown in Fig. 14.1.

The new method is based on the following:

1. use of activated charcoal as a carrier material, an absorption base, and for maintaining anaerobiosis
2. use of most effective protective agents
3. two-steps primary-drying under mild vacuum
4. maintenance of process temperature at 29°C
5. short secondary drying for double vial preparations.

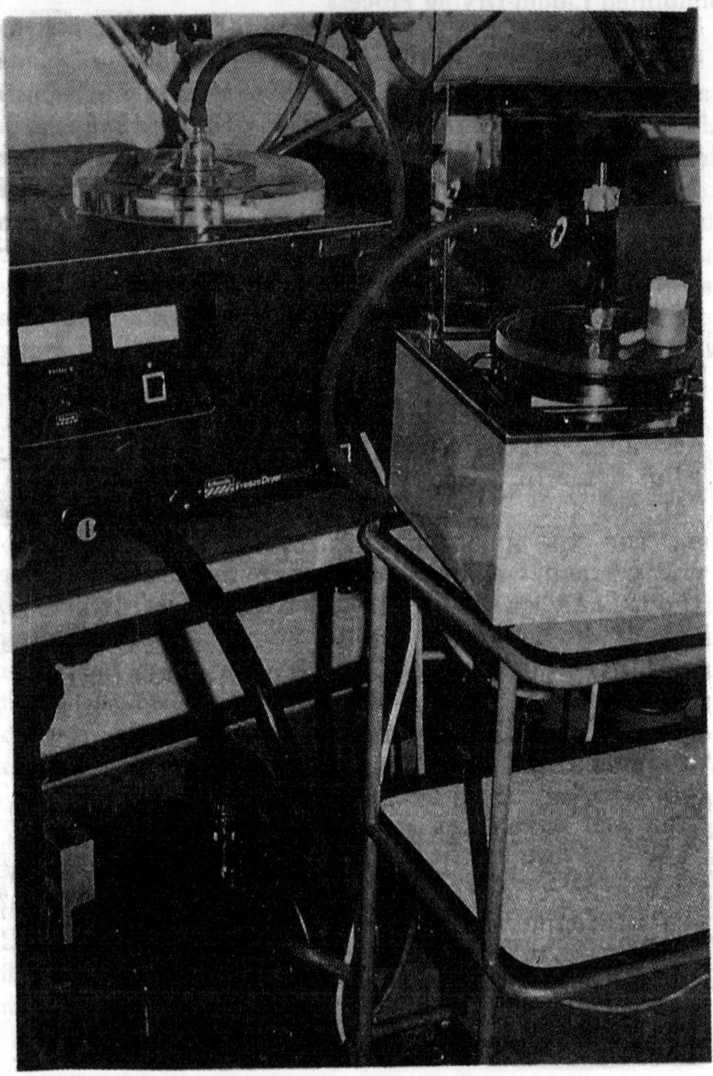

Fig.14.1: Preservation by liquid-drying at the DSM

Liquid-drying of Micro-organisms Under Anaerobic Conditions

A new liquid-drying method for the long-term preservation of micro-organisms under anaerobic conditions has recently been described (Malik, 1991c). In this method, drops of cell suspensions are dried under mild vacuum on thin discs of activated charcoal containing effective protective agents. Gas-tight small screw-cap glass vials with rubber septa are used for drying the specimen anaerobically. No constriction or heat sealing of glass vials is needed. During experimental manipulations, no anaerobic chamber or glove boxes are required. The new method has successfully been applied for the preservation of a collection of sensitive and anaerobic micro-organisms. The method is simple, effective, economical, and is also suitable for drying of micro-organisms in small laboratories. The outline of this liquid drying procedure and the major steps involved are shown in Fig. 14.2.

During L-drying under reduced conditions, activated charcoal proved to be a neutral and harmless reducing agent especially for such cultures, which are sensitive to other toxic reducing agents. Several anoxygenic light sensitive phototrophic bacteria have successfully been dried in the presence of activated charcoal which by absorbing light and oxygen proved effective against photo-oxidation.

Although the use of effective protective agents increases the survival and stability of micro-organisms during L-drying process, harmful effects due to the presence of various such additives (dimethylsulphoxide, thiourea, cystein, sodium sulphide, etc.) in the rehydration medium during the reactivation process have been reported. A suitable rehydration medium during reactivation of dried cultures improves survival and is reported to be involved in the repair mechanism of DNA damage that results during drying. The presence of activated charcoal in the dehydration medium proved not only harmless but also helpful during reactivation by absorbing various harmful radicals (superoxide and other free radicals created by various electron donors in reaction with oxygen).

Liquid-drying of Micro-organisms Using Simple Apparatus

In this method, liquid-drying can be carried out without the use of freeze-drying machines or other specialized equipment (Malik, 1991a). Throughout the procedure the temperature of the cell material is maintained at 20°C.

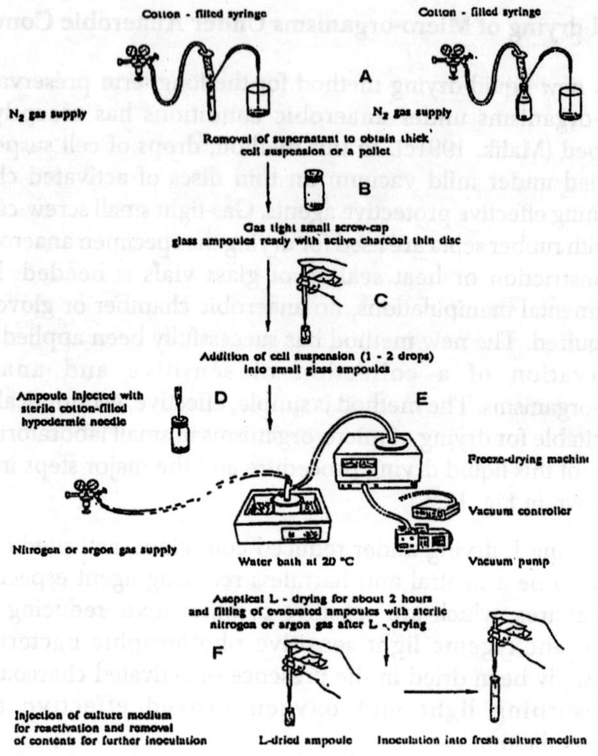

Fig. 14.2: Liquid-drying of micro-organisms under anaerobic conditions

This methods is based on a new simplified L-drying method which has been described above (Malik, 1990a). It is simple and the equipment described here can be easily constructed in most laboratories.

An outline of the new liquid-drying procedure using simple apparatus and the major steps involved have already been give in the course manual. More details on this method are described in Malik (1991a).

FREEZE-DRYING (LYOPHILIZATION)

Freeze-drying is often used for the preservation and long-term storage of bacteria, fungi, yeasts and viruses and it is probably the most widely used technique. Starting from very simple equipment of

Vansteenberghe, used in 1903, advanced freeze-drying machines have now been developed. Sophisticated freeze-drying machines have three major components: a freezing mechanism, a vacuum attachment and a water trap (Fig. 14.3).

Freeze drying is commonly achieved by a prefreezing method or by a centrifugal freeze-drying method. In the prefreezing method, samples are frozen outside the machine and then transferred to the vacuum chamber without thawing.

Centrifugal freeze-drying involves the centrifugation of the suspensions and cooling by evaporation after the application of vacuum. Once the material is frozen, the water is removed from it to achieve a suitable final residual moisture content.

During this stage, usually overdrying results, which may harm the cells. This can be prevented by using a suitable suspending media since during unprotected freeze-drying, large numbers of cells may be killed or undergo mutations.

This has already been discussed in detail and the precautions to take to reduce freeze-dry damage have been summarized (Malik, 1987, 1988b).

During the freeze-drying process, microbial suspension is frozen, and water is quickly removed by sublimation directly from ice. The process ends with a dried product which, if sealed either under vacuum or under inert gas, can be stored at room temperature with no metabolic activity possible until water and nutrients are restored. The cultures are thus viable, but metabolically inactive which can easily be returned to active growing conditions.

New Freeze-drying Method for Micro-organisms

Optimizing of the Freeze-drying Protocol

Freeze-drying is a widely used technique and there are many variations in technical details. Generally it is achieved by rapidly freezing and subsequent drying. Durying comparative studies, it has been observed that even good additives proved less effective during this process as compared to slower freezing of the samples prior to freeze-drying.

Considering this and several other parameters, a new freeze-drying method has recently been reported in which technical

Fig. 14.3: Freeze-drying appartus used at the DSM

details have been defined and optimized for the preservation of a comprehensive collection of nitrogen-fixing and other fragile bacteria (Malik, 1988a). The method is based on slow freezing of cell suspension (about 1°C per min to about-30°C) prior to prolonged primary freeze-drying (with slow sublimation). This is followed by short (2-3 hours) secondary drying. The outline of the new freeze-drying procedure using standard apparatus and the major steps involved are shown in Fig. 14.3. More details on this method are described in the course manual.

This method provides a suitable model for freeze-drying of many delicate strains and is applicable to a broad range of micro-organisms.

Freeze-drying of Micro-organisms Under Anaerobic Conditions

Freeze-drying of anaerobes is often difficult. Several anaerobic bacteria which are sensitive to aerobic freeze-drying, can now be preserved successfully using activated charcoal (5%w/v) in the suspending media along with the above protective agents without the use of glove box or an anaerobic chamber (Malik, 1990b). However, for freeze-drying of very sensitive anaerobes, standard anaerobic conditions are maintained prior to freeze-drying. The details of the freeze-drying procedure and the major steps involved are the same as described above in the new freeze-drying method.

Freeze-drying of Micro-organisms Using Simple Apparatus

Where there is access to a freeze-dryer, simple freeze-drying can be achieved with an evacuation, desiccator, a condensor, and a vacuum pump (fig. 14.4). Pre-frozen ampoules are then placed in the desiccator (with a desiccant like P205) and then evacuated with a pump. A new freeze-drying method has recently been published which can be carried out without the use of freeze-drying machines or other specialized equipment (Malik, 1990c). The method is simple and the equipment described here can be easily constructed in most laboratories. The outline of the new freeze-drying procedure using simple apparatus and the major steps involved have already been given in (Malik, 1990c).

DEEP-FREEZING IN LIQUID NITROGEN (CRYOPRESERVATION)

Freezing provides another possibility of short-term storage. During simple freezing, bacterial thick suspensions are frozen at low temperature of about-30°C which can easily be attained through commercially available deep-freezers. Although some microbes can be stored temporarily at such temperatures, most of these are killed due to damage during the freezing process. Freezing of cultures at temperatures below-70°C is more suitable. As temperature is lowered, the growth of ice crystals diminishes until somewhere below-130°C the ice crystal growth is essentially negligible. Thus storage below-130°C assures long 'shelf-life', though for long-term storage, the best temperature is attained by the use of liquid nitrogen (-196°C). Cryopreservation has proved to be a valuable technique for the long-term preservation of cells sensitive to simple freezing and drying, and it is the preferred method for almost all biological materials. Liquid nitrogen is non burning, non-toxic, safe, and relatively cheap compared with other gases.

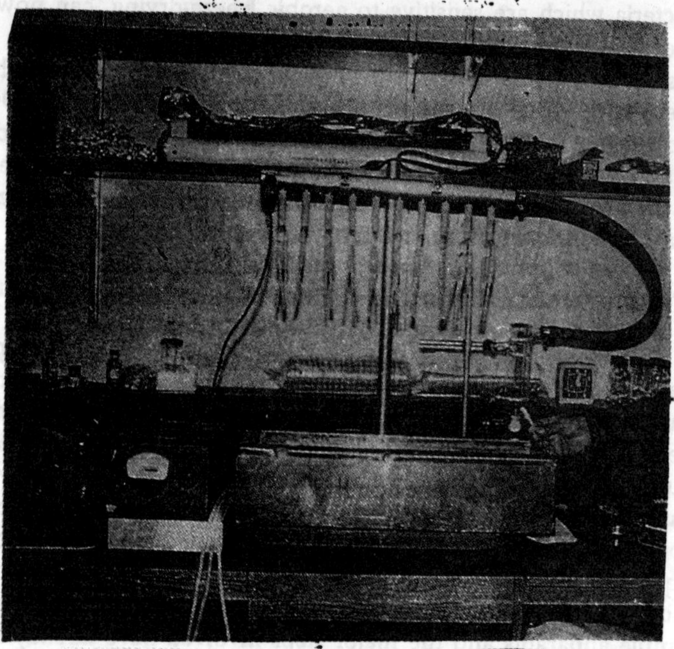

Fig. 14.4: A simplified freeze-drying apparatus used at the North Regional Researc Centre in the USA.

During freezing, internal drying in cells begins at the moment when ice begins to form in the external medium and is essentially complete at-196°C, when drying is defined on the basis of available water. Thus the greater the rate of freezing, the slighter the movement of water from the cells to the medium, and the greater the chance of formation of internal cellular ice. This does not appear to be as significant for small cells as for cells in the size range of the yeasts, where this could be critical. Moreover, as water is effectively removed during freezing as ice, electrolytes become increasingly concentrated in unforzen water, which is harmful: electrolyte concentrations outside cells are very different from those inside cells. With liquid nitrogen storage, the rate of cooling is thus important, which differs for different biological materials. Generally, a rate of 1-10°C/min (for yeasts 10°C/min and for bacteria 1-5°c/min) provides a satisfactory recovery for micro-organisms. The greater efficiency in recovery of microbial cells from the frozen state is important because it minimizes the possibility of the selection of cells resistant to the technical manipulations employed during the process of cryopreservation.

With the use of commercially available programmable freezers, appropriate cooling rate can easily be achieved. In practice a relatively slow initial cooling rate can also be obtained by keeping cultures in the neck of the liquid nitrogen storage unit for some minutes and then lowering containers into it. It is not, however, good practice to plunge cultures directly into liquid nitrogen. Indeed, for other safety reasons, the cultures should be stored in vapour phase above the liquid nitrogen itself. The safety aspect arises because if immersed, then liquid nitrogen will seep into any imperfectly sealed capillaries, ampoules, or vials that contain the cell suspensions: on removal from storage, this nitrogen will become gaseous phase, almost instantly causing an explosion.

It is generally found that best recovery of cultures from liquid nitrogen is achieved by quick thawing. However, the events during thawing are complex and it is not only a simple reversion of the freezing process. Slow warming rates can lead to ice crystal formation whereas a rapid warming at about 100°C per min generally will give good survival rates. To achieve this, usually the samples (cryovials) removed from liquid nitrogen are quickly immersed into water baths at 37°C.

With controlled freezing and thawing it is possible to freeze-store microbes with good genetic stability and viability (Simione *et al.*, 197; Daily and Higgens 1973; Kirsop and Snell, 1984; Malik, 1984). For protection of cells from freeze-damage, cryoprotective agents should be used as only very few cell types will survive freezing in liquid nitrogen without these. Compounds which protect living cells against freeze-thaw injury are termed cryoprotectants and belong to two main classes on the basis of their permeability to cells. The penetrating compounds such as dimethylsulphoxide (DMSO), glycerol, methanol, and ethylene glycol are low molecular weight compounds and are generally applied in high concentration (0.5-1.5 molar). The non-penetrating agents such as hydroxy-ethyl (HES), polyvinyl pyrolidone (PVP), sugars, proteins, and polyethylene glycole are used at much lower concentrations (0.01 molar). Toxicity of these compounds at higher concentrations may cause problems. All these substances have common important properties: their high affinity for water and solubility even at low temperatures. This may relate to their protective ability, but their exact mode of action is not known and may be as diverse as their structure.

In liquid-nitrogen storage of micro-organisms use is generally made of polypropylene cryotubes, glass vials, glass capillaries, and polypropylene straws (Kirsop and Doyle, 1991). Usually, for effective cryopreservation of anaerobic micro-organisms, experimental manipulations are performed within an anaerobic chamber or use of cumbersome methods, like heat sealing of glass ampoules under a stream of nitrogen gas. A new method has been published (Malik, 1984) for liquid nitrogen storage of phototrophic bacteria under anaerobic conditions. In this method, mini screw-cap glass ampoules were used. These are suitable for preserving bacteria under anaerobic conditions. The method is simple, effective and economical with respect to the storage space and costs. It was specially designed for the cryopreservation of anaerobes. In the preservation of anaerobic phototrophic or other bacteria, a continuous stream of nitrogen gas, anaerobic chamber, and glove boxes are not generally required. Almost all existing species of phototrophic bacteria at the DSM have been successfully preserved with this method and it provides a suitable model for cryogenic storage of many sensitive and delicate micro-organisms. The method is equally good for the economical preservation of aerobic micro-organisms. Details of cryopreservation under anaerobic conditions are given in the course manual–see also Fig. 14.5.

PRESERVATION OF MICRO-ORGANISMS BY SIMPLE METHODS

The long-term preservation of cultures in small laboratories is often very difficult owing to the absence of special equipment for deep-freezing or freeze-drying The storage and preservation of cultures using simple methods is however, possible in any laboratory (Kirsop and Snell, 1984; Malik, 1985, 1991a). In the course manual, practical details on drying and freezing of micro-organisms on glass or porcelain beads are described. These techniques can be carried out in any microbiological laboratory and for many years these have been demonstrated and practically performed in various preservation workshops and training courses (Malik, 1991a). The methods have proved to be safe, reliable and effective for the storage of a wide range of micro-organisms. However, viability, shelf-life and the stability of cultures is relatively poor when compared wi h the preservation in liquid nitrogen and liquid-drying (Malik, 1987).

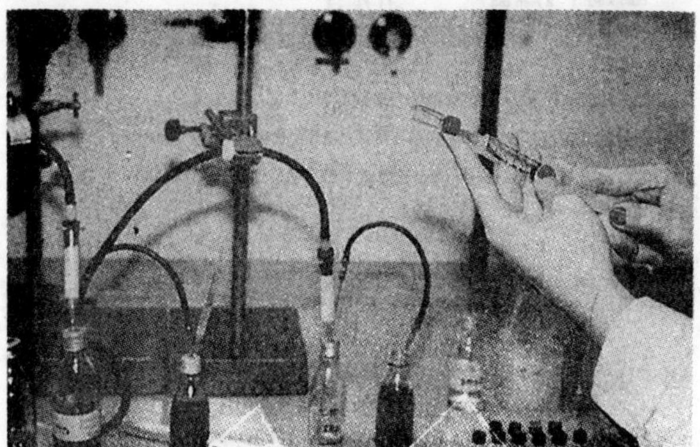

Fig. 14.5: Cryopreservation in liquid nitrogen under anaerobic conditions.

Micro-organism Drying on Glass or Porcelain Beads

Cultures of several micro-organisms that are sensitive to freezing or freeze-drying, can be preserved successfully by drying by application of a more delicate treatment (Malik, 1991). In simple drying procedures, a small volume of cell suspension is spread over large surface areas. In this method, cells are spread over the surface of glass or porcelain beads and a mild drying is achieved over a desiccant without applying vaccum. It is a very useful method especially for spore-forming fungi and bacteria. The average storage period for spore-forming micro-organisms is over ten years. With the use of a protective agent (milk, honey, sugars, serum and others), even non-spore-forming bacteria can successfully be maintained for many years. The technical details of the method are described in detail in the course manual.

Micro-organism Cryopreservation on Glass Beads

There are several simple and economical methods available for cryopreserving micro-organisms. The method of freezing micro-organisms on glass beads is very simple and effective for the storage of a wide range of micro-organisms (Kirsop and Doyle, 1991). Very little material is needed and many strains can be stored for long periods in a relatively small space. Culture recovery is quick, and meanwhile disturbance to other stored specimens is almost entirely avoided. Details of this method are described in the course manual; the major steps involved are shown in Fig. 14.6.

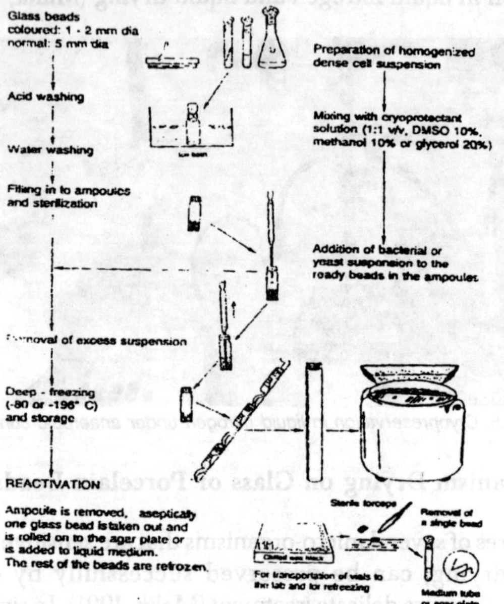

Fig. 14.6: Cryopreservation of micro-organisms on glass beads.

REFERENCES

Annear, D.I. 1962. Recoveries of bacteria after drying on cellulose fibres. *Austral. J. Biology and Medical Science*, 40, 1-8.

Banno, T. and Sakane, T. 1979. Viability of various bacteria after L-drying. *IFO Research Commun.*, 9, 35-45.

Daily, W.A. and Higgens, C.E. 1973. Preservation and storage of micro-organisms in the gas phase of liquid nitrogen. *Cryobiology*, 10, 364-67.

Greiff, D., Melton, H. and Rowe, T.W. 1975. On the sealing of gas-filled glass ampoules. *Cryobiology*, 12, 1-14.

Heckly, R.J. 1985. Principles of preserving bacteria by freeze-drying. *Developments in Industrial Microbiol.*, 26, 379-95.

Kirsop, B.E. and Doyle, A. (eds). 1991. *Maintenance of Micro-organisms and Cultures Cells.* Academic Press, London.

Kirsop, B.E. and Snell, J.J.S. (eds) 1984. Maintenance of Micro-organisms: *A Manual of Laboratory Methods.* Academic Press, London.

Mackenzie, A.P. 1977. Comparative studies on the freeze-drying survival of various bacteria: gram-type, suspending medium and freezing rate. *Dev. Biol. Standard* (Karger, Basel), 36, 63-277.

Malik, K.A. 1976. In H. Dellway (ed), *Proc. Fifth Int. Fermentation Symp.* Westkreuz Druckerei and Verlag, Bonn and Berlin. pp. 180.

Malik, K.A. 1977. Rapid surface colony counts determination with three new miniaturised techniques. *Zbl. Bak. Hyg. I. Abt. A*, 237, 415-23.

Malik, K.A. 1984. A new method for liquid nitrogen storage of phototrophic bacteria under anaerobic conditions. *J. Microbiol. Methods*, 2, 41-47.

Malik, K.A. 1985. *Modern Methods of Gene Conservation. A Laboratory Manual.* ASTIC Press, Pakistan S & T Information Centre, Islamabad.

Malik, K.A. 1987. In J. Amen and P. Tesson (eds), *The Role of Culture Collections in the Stability and Preservation of Micro-organisms.* Societe Francaise de Microbiologie, Paris.

Malik, K.A. 1988a. A new freeze-drying method for the preservation of nitrogen-fixing and other fragile bacteria. *J. Microbiol. Methods*, 259-71.

Malik, K.A. 1988b. Long-term preservation of some Rhodospirillaceae by freeze-drying. *J. Microbiol. Methods*, 8. 273-80.

Malik, K.A. 1990a. A simplified liquid-drying method for the preservation of micro-organisms sensitive to freezing and freeze-drying. *Microbiol. Methods*, 12, 125-32.

Malik, K.A. 1990b. Use of activated charcoal for the preservation of anaerobic phototrophic and their sensitive bacteria by freeze-drying. J. *Microbiol. Methods*, 12, 117-24.

Malik, K.A. 1990c. Freeze-drying of micro-organisms using a simple apparatus. WFCC Technical Information Sheet. UNESCO/WFCC-Education Committee Publication No. 7.

Malik, K.A. 1990d. Liquid-drying of micro organisms using a simple apparatus. WFCC Technical Information Sheet. UNESCO/WFCC-Education Committee Publication No. 8.

Malik, K.A. 1991a. Maintenance of micro-organisms by simple methods. In B.E. Kirsop and A. Doyle (eds), *Maintenance of Phototrophic Bacteria. Maintenance of Micro-organisms and Cultured Cells.* Academic Press, London.

Malik, K.A. 1991c. A new method for liquid-drying of micro-organisms under anaerobic conditions. *J. Microbiol. Methods.* In press.

Malik, K.A. 1991d. In B.E. Kirsop and A. Doyle (eds), *Maintenance of Micro-organisms.* Academic Press. london. pp. 82-100.

Malik, K.A. 1991 e. Long-term preservation of green algae by liquid-drying. Manuscript in preparation.

Simione, F.P., Daggett., P.M., MacGrath, M.S. and Alexander, M.T. 1977. The use of plastic ampoules for freeze preservation of micro-organisms. *Cryobiology*, 14, 500-02.

15

COLLECTION AND CRYOPRESERVATION OF ANIMAL CELLS

C.B. MORRIS
ECACC, PHLS-CAMR, Porton Down, Salisbury, England

INTRODUCTION

Animal cell culture collections are one of the more recent additions to the expanding world of culture collections for living organisms. The technique of culturing animal and human cells originates from the turn of this century when attempts to understand the physiological role of tissues were made by producing explant cultures in simple salt solutions (Harrison, 1907).

Although the development of culture techniques gradually improved with the use of biological fluids to support cell growth, no readily available storage methods could be found. It was not until the production of liquefied gases that the next development occurred.

Attempts to cryopreserve animal cells by placing them in gaseous or liquid nitrogen were largely unsuccessful owing to the loss of cell viability on reduction. This technique, however, was extremely successful with micro-organisms such as bacteria.

The need for inclusion of a cryoprotectant with the cells was realized when Polge (1949) discovered that addition of glycerol to spermatozoa enhanced survival during cryopreservation. Later other diverse chemicals such as dimethyl pyrolidone (PVP) were also found to provide protection against freezing.

THE MECHANISM OF CRYOPRESERVATION

The mechanisms involved in cryopreservation were elucidated by studying the effects of varying the cooling rate and warming rate on cell viability. Optimum survival generally required that the cells are cooled at a slow rate, i.e.,-1°C/min to-5°C/min and then on resuscitation thawed rapidly to the normal growing temperature i.e., in a water bath (Mazur, 1977).

Studies with two-step freezers illustrate the fundamental mechanisms involved. This apparatus consists of a cryogenic dewar with a long neck. The main body of the dewar is filled with liquid nitrogen but not into the neck. After placing on the cap the vessel is allowed to equilibrate.

A canister containing ampoules of cells for freezing is placed inside the neck in the vapour phase. After a holding time of 10-20 minutes the canister is plunged into the liquid and then transferred to a storage vessel.

Ideally, a position in the gas phase that provides a holding temperature of -20°C to -25°C gives maximum survival, although some cell viability is usually lost during freezing.

The requirement for holding the cells at a constant temperature is necessary for two reasons. This allows time for water to leave the cells via the plasma membrane with the concomitant result that the cells shrink, reducing the amount of water in the cell that can freeze at subzero temperatures. As ice formation is the major form of damage occurring during freezing, causing physical disruption of internal membranes and organelles, optimum survival can only be achieved if the water content is minimal. (Hunt *et al.*, 1977; Griffiths and Beldon, 1978; Clegg *et al.*, 1982.)

During slow cooling, aqueous solutions supercool a few degrees below their freezing point before the water begins to freeze. This is a relatively inefficient process which is initiated by nucleation of foreign particles in water molecules acting as a focal point for ice crystal formation. Once a crystal forms it grows, reducing the water content of the solution. As the solutes increase in concentration, a further temperature drop is required to convert more water to ice.

The intracellular components providing nucleators for ice formation are even less efficient than those in the extracellular solution.

Consequently, this creates an osmotic imbalance across the plasma membrane as the extracellular solute concentration is always higher than the intracellular concentration at any given temperature above that at which all the water is frozen. A schematic illustration of this is shown in Fig. 15.1.

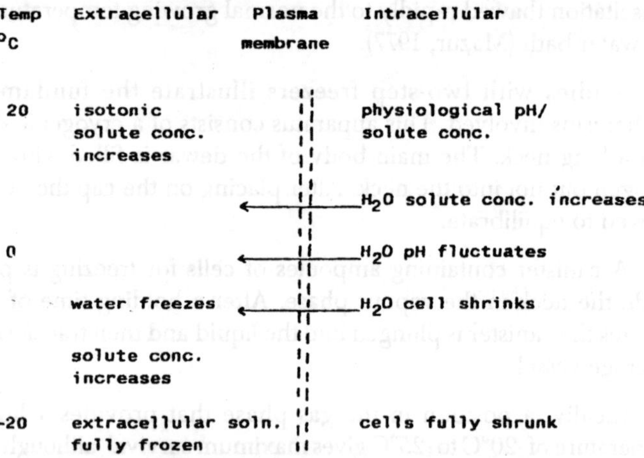

Fig. 15.1. The effect of slow cooling on solute concentration.

During cooling, water in the extracellular solution freezes, raising the solute concentration and causing an osmotic imbalance to occur across the plasma membrane.

As the plasma membrane is semipermeable, it allows free passage of water, which means that osmotic equilibrium is restored by an efflux of water from the cells. The rate at which water can leave the cells is determined by the osmotic gradient and the membrane characteristics of individual cell lines. In practice, this means that the optimum cooling rate will vary for different cell lines.

The addition of a cryoprotectant to the freezing medium does not mitigate these effects, but it does allow the cooling rate to be reduced. This is due to a suppression of the freezing point, thus allowing further supercooling of the soluble components.

By increasing the time taken to reach total cell freezing, more water can exit from the cells to achieve a maximum intracellular water loss.

The cryoptotectants most commonly used with animal cell lines are DMSO and glycerol, both of which penetrate the cells. It should be remembered, however, that cryoprotectants such as DMSO can be toxic to cells during prolonged exposure.

Several factors play a role in achieving optimum cryopreservation, further discussion of which is given below.

WATER PERMEABILITY

Both the osmotic gradient and membrane characteristics affect the rate of water movement across the membrane. Therefore, as the rate of water efflux increases, the rate of cooling can be increased.

A consequence of the increasing solute concentration can be abrupt changes in the intracellular and extracellular pH, in particular calcium, magnesium and hydroxyl ions. This can lead to localized damage of membranes and protein precipitation.

CELL SIZE

Small cells have a higher surface area to volume ratio than large cells, which not only allows them to potentially lose water at a faster rate, but in practice they can shrink to a smaller volume, relative to their original volume.

CELL CONCENTRATION

Each cell type has an optimum density range for survival during cryopreservation. If this is exceeded the ratio of the total freeze mixture and total cell volumes, decreases to a point where there is insufficient extracellular solute to achieve maximum water loss from the cells. A density of between 3-10 × 10^5 cells/ml is normally used.

TYPE OF CRYOPROTECTANT

The decision to use a penetrating, e.g., DMSO, glycerol or non-penetrating, e.g., dextran, PVP will depend on the cell type being frozen. The normal concentration range used is between 5-10% of the final volume. DMSO may cause increases in pH, especially if media containing bicarbonate are included in the freeze medium.

ADDITION OF CRYOPROTECTANT

When cells are rapidly exposed to cryoprotectants an osmotic imbalance can result owing to the higher solute concentration outside the cell.

This can be overcome by either a slow addition to the cells, or allowing time for the cells to reach an osmotic equilibrium after adding the freeze medium, i.e., 5-10 minutes.

METHOD OF COOLING

Slow cooling is essential to achieve maximum survival. Ideally the rate of cooling should be constant, something that can only be attained with a programmable rate freezer.

The practice of using insulated containers, e.g., polystyrene, placed in -70°C freezers will give varying degrees of success depending on how well the final cooling rate is regulated. This can only be tested by trial and error with different amounts of insulation.

The two-step freezer is more reproducible but also does not overcome the need for increased cooling during the phase transition period when water is freezing.

A property of water is that as it converts from one state to another, i.e., liquid to solid, latent heat of fusion occurs with a resultant rise in temperature. This can only be overcome by temporarily increasing the cooling rate at this point.

Depending on composition, freezing mediums will usually freeze between-4°C to -8°C. Programmable freeze can increase the cooling rate to correspond with this freezing point and consequently eliminate the accompanying rise of temperature that would normally occur.

METHOD OF THAWING

With few exceptions, frozen cells must be thawed rapidly to their normal growing temperature, e.g.,+25°C (insect cell lines), + 37°C (mammalian cell lines). This can be best achieved using a water bath.

Slow thawing will increase the damage caused by ice crystals and longer exposure to osmotic imbalance.

After thawing, cells will be in a more fragile state than growing cultures. Therefore should centrifugation be required it must be at low speed and for short periods to minimize additional stress, e.g., shearing forces.

STORAGE TEMPERATURE

Temperatures above-130°C are not suitable for long-term storage. In instances when-70°C freezers have been used to store animal cells, a rapid decrease in cell viability has been observed over a period of a few months.

Ideally, the gas (-140°C) or liquid (-196°C) phase of liquid nitrogen should be used.

PRACTICAL ASPECTS OF FREEZING ANIMAL CELLS

A list of equipment and material requirements is shown in an accompanying training manual available from our group (Practical 3).

When starting a cell line from original tissue, a cell banking procedure should be initiated as shown on P3-1 of the training manual. Preparation of cell banks in this manner provides total control over cell lineage and records the number of subculturings (passages) or population doublings the cells have undergone at the time of preparing a working bank.

Quality control checks at each of the freezing stages, i.e., token, master and working banks, are essential to ensure the stock is free of microbial contamination, i.e., bacteria, fungi and mycoplasma (McGarrity, 1982; Mowles, 1990), and have been characterized for authenticity, i.e., karyotyping, isoenzyme analysis and DNA fingerprinting (O'Brien *et al.*, 1981; Stacey *et al.*, 1991).

Cultures initiated from the working bank should only be maintained for a set number of passages of time period, to reduce the possibility of genetic drift, microbial or viral contamination and cell line cross-contamination. The problems arising from this last point are best illustrated by a study undertaken by Nelson-Rees *et al.* (1981).

The authenticity of a cell line's species of origin can readily be checked using isoenzyme analysis. A kit for this purpose is obtainable from Innovative Chemistry.* Extracts from cells are run on agarose gels against a known standard and control using up to seven enzymes for identification. Comparison of results with those for known species allows rapid identification of the test material. The kit is very easy to use and the test can be completed in a few hours.

A detailed method of the preparation of animal cells is given elsewhere. However, there are additional points for consideration.

STARTING TEMPERATURE OF FREEZE RUN (PROGRAMMABLE FREEZER)

A discrepancy between the temperature of the prepared cells and the chamber temperature, e.g., cells at +15°C, chamber at +4°C, will mean that the programme of a freeze run initiated with this temperature difference will be running in advance of the cell temperature, i.e., the increase in cooling rate (to overcome the heat generated at phase transition) will come too soon, and when the cells finally begin to freeze, no increase in cooling rate will occur.

The cells and the chamber starting temperatures must be the same. This problem can be overcome by setting the programme to hold at +5°C for 10 minutes prior to starting the run.

FREEZE MEDIUM COMPOSITION

Traditionally this is prepared from the cells growth medium with an elevated level of serum, i.e., 20-25% v/v and 10% cryoprotectant. However, many cell culture media contain bicarbonate as a buffering agent. Addition of DMSO and the increasing concentration of the bicarbonate ions during freezing will raise the pH above the normal physiological tolerance range of the cells.

This can be overcome by using whole serum, e.g., FBS or NBCS with cryoprotectant (8-10%). This mixture has the added advantages of providing extra cryoprotective action from the presence of serum albumins, and neutralization of proteolytic enzymes used to remove attached cells.

---

* Authentikit from Innovative Chemistry Inc., PO Box 90, Marshfield, MA 02050, USA.

It is advisable to select serum batches for growing as well as freezing cell lines, carefully. The source, country or origin and quality control tests carried out on a batch of serum can have serious implications for the future use of a cell line, especially if it is to be used in a commercial process, or shipped to another country, e.g., USA, Japan.

Only serum from reputable sources should be used with cell lines and if they are to be sent to other countries, the import requirements should be checked in case serum sourced from certain countries, e.g., South America, cannot be imported because of the risk of banned viruses being present in the serum.

TYPE OF CRYOPROTECTANT

Although DMSO is a most widely used cryoprotectant, this does not necessarily make it the one choice for all cell lines. Some cell lines spontaneously differentiate in its presence, e.g., HL-60, a human promyelocytic leukaemia, and in the experience of the author, mouse T-cells survive better when glycerol is used.

If a cell line has specialized properties, these should be checked after freezing to ensure stable supplies for the future.

SERUM/PROTEIN-FREE CULTURES

If a cell line is grown in a serum or protein-free media, it may be necessary to freeze the cells in the absence of serum or protein. Experience has shown that the total removal of protein from a freezing medium can cause a dramatic fall in cell survival, even when a programmable freezer is used.

The brief exposure to serum during cryopreservation may not change the cell growth behaviour. However, it may have implications for regulatory purposes if the possible presence of viruses in serum compromises any product produced by the cells, e.g., monoclonal antibodies.

MAINTENANCE OF RECORDS

Accurate details of all cell lines frozen including source of the line, freeze medium, passage number, storage locations, and date of preparation must be kept for future reference. A plan of storage locations can be supplemented by the use of databases.

Computer software for storage documentation can be obtained from Business Simulations Ltd. ('Cardbox Plus'–a non-relational system) and Ashton-Tate ('dBaseIV'–a relational system).

REFERENCES

Clegg, J.S., Seitz, P., Seitz, W. and Hazelwood, C.F. 1982. Cellular responses to extreme water loss: the water replacement hypothesis. *Cryobiology,* 19, 306.

Griffiths, J.B. and Beldon, I. 1978. Assessment of cellular damage during cooling to-190° using radiochemical markers. *Cryobiology,* 15, 391.

Harrison, R.G. 1907. Observations on the living developing nerve fibre. *Proc. Soc. Exp. Bio. Med.,* 4, 140-43.

Hunt, C.J., Beadle, D.J. and Harris, L.W. 1977. An ultrastructural study of the recovery of Chinese hamster ovary cells after freezing and thawing. *Cryobiology,* 14, 135.

Mazur, P. 1977. The role of intracellular freezing in the death of cells cooled at supraoptimal rates. *Cryobiology,* 14, 251.

McGarrity, G.J. 1982. Detection of mycoplasmal infection of cell cultures. In K. Maramorosch (ed), *Advances in Cell Cultures,* Academic Press. pp. 99-131.

Mowles, J.M. 1990. Mycoplasma detection. In Pollard and Walker (eds), *Methods in Molecular Biology. Vol. 5: Animal Cell Cultures.* Humana Press Inc.

Nelson-Rees, W., Daniels, D. and Flandermyer, R. 1981. Cross-contamination of cells in culture. *Science,* 212, 446.

O'Brien, S., Shannon, J. and Gail, M. 1981. A molecular approach to the identification and individualisation of human and animal cells in culture: isozyme and allozyme genetic signatures. *In Vitro,* 16, 119.

Polge, C., Smith, A.N. and Parkes, A.S. 1949. Revival of spermatozoa after vitrification and dehydration at low temperatures. *Nature,* 164, 666.

Stacey, G.N., Bolton, B.J. and Doyle, A. 1991. The quality control of cell banks using DNA fingerprinting: approaches and applications. In Burke, Dolf, Jeffrey and Wolff (eds), *DNA Fingerprinting.* Birkhauser Verlag, Basel. pp. 361-70.

16

COLLECTION, IDENTIFICATION AND PRESERVATION OF FUNGI AND YEASTS

P. HOFFMANN
DSM—German Collection of Micro- organisms and Cell Cultures, Braunschweig, Germany

INTRODUCTION

The number of known fungi is enormous in comparison to other micro-organisms. About 75,000 species are currently described, with 1500 species being added every year according to the IMI Index of fungi. The actual number of species is estimated to be around 250,000 (Hawksworth, 1988), but only several thousand have been successfully grown in pure culture and exciting new and interesting species with novel properties still wait to be detected.

ISOLATION

Fungi play a major role in the recycling of organic matter and can be found almost everywhere on earth. Many are saprophytic, degrading complex substrates, but some live as parasites in or on other organisms or as symbiotic partners for mutual benefit, like the mycorrhiza-forming fungi. The isolation of these species into a pure culture is essential for their exploitation in science and technology.

Before any isolation is attempted, the objective of the programme should be clearly defined. The choice of isolation techniques is determined by the purpose of the study: enumeration of all fungi present in a certain habitat, isolation of strains with special physiological characteristics, or screening for the production of metabolites require a different set of media and incubation conditions. Generally, it can be said that the conditions for isolation should reflect the ecology of the species: xerophilic species need media of low water activity, thermophilic fungi should be incubated at 40-45°C. Detailed descriptions of media and methods are given by Booth (1971).

In contrast to bacteriology, special selective media play only a minor role in mycology, though to some extent all isolation methods are selective. By opening a Petri dish in the laboratory, for instance, only the airborne spores will be trapped. Some groups of fungi can be inhibited or promoted by the addition of special chemicals to nutrient media. A compilation by Gams *et al.* (1987) lists antibiotics like chloramphenicol or tetracylin to suppress bacterial contaminants; oxgall or sorbose to inhibit fast-growing species so that slow-growing fungi can be detected; benomyl and o-phenylphenol help to suppress 'weed fungi' during the isolation of basidiomycetes; ethanol in low concentration will inhibit many yeasts except the strongly fermenting ones and some wood-roting species (e.g., *Heterobasidion annosum*); high amounts of tannin select in favour of tannase-producing species like *Aspergillus niger;* creosote will suppress all fungi in soft probes except *Hormoconis resinae.* Heating of the substrate (80°C, 30-60 min) will allow the selective isolation of heat-resistant cleistothecial ascomycetes (*Eupenicillium, Byssochlamys,* etc.).

Numerous techniques are in use for the isolation of fungi from different substrates like soil, plan and animal tissue, leaf surface, and water. The simplest is the direct isolation of spores and fruiting structures. Mould conidia, asco- and basidiospores of wood-rotting fungi generally germinate and grow on standard media like malt extract, cornmeal, or potato dextrose agar. These spores can be picked directly from fruiting structures with the aid of a low-power microscope and fine needles or they can be collected in water, which is diluted until a sufficiently low density of spores is availabie for planting. From the carpophores of fleshy fungi, cultures can be obtained by aseptic regeneration of tissue. Symbiotic or parasitic fungi can be isolated from host tissue incubated in a humid chamber.

Sterilization of outer surfaces usually precedes attempts to isolate fungi from their hosts. The type of sterilizing agent, its concentration, time of treatment, etc., largely depend on the sensitivity of the fungus wanted. Widely used are ethanol (75%, 1-5 min), hypochlorite solution (10%, 1-4 min, extensive washing afterwards), or mercuric chloride (0.1%, 2-4 min, washing for 30 min).

Dormancy or inhibited germination can be found in asco- or basidiospores from certain species, e.g., many fungi growing on dung. No universal treatment is available to overcome this phenomenon. If it can be attributed to an inhibitor, washing the spores or diluting the spore suspension may help. Sometimes a mild heat shock (40°C, 20 min) will improve the rate of germination. In many basidiomycetes, the presence of mycelium of the same of closely related species is necessary for basidiospore germination. Colonies of red yeast and activated charcol may have a similar effect– the same has been reported for chemicals like glycocholate or furfural. The same techniques used for isolation can be applied to the purification of contaminated cultures.

IDENTIFICATION

The process of fitting an unknown culture into the existing framework of taxonomical groups is called identification. Accurate identification is the key to understanding ecological relations of species and exploiting their potential for technological use. Traditionally, the morphology of sexual and asexual fruiting structures has been used for this purpose, but in relatively complex genera like *Penicillium* more and more physiological and genetical characteristics have been introduced to give a reliable classification without too much subjectivity. In morphologically simple groups like the yeasts, physiological data, such as the ability to grow on various carbon and nitrogen sources, have always played a major role in classification and identification procedures.

The prerequisites for identification are a pure culture in an optimal developmental stage, knowledge of mycological terminology, and an identification key. Contaminated, degenerate or even atypical strains may be difficult or impossible to identify. For many groups of fungi, the literature is sparse, and complete modern keys are lacking; original descriptions may only help sometimes. Keys to major groups of fungi are given by Ainsworth *et al.* (1973), a key to the genera of fungi sporulating in pure culture has been published by von Arx (1981), both giving a guide to more literature.

There is no comprehensive list of characteristics important for identification purposes. Whereas for the description of a species as much information as possible should be obtained, identification usually relies on a small set of 'key characters', which, however, may vary even between closely related genera. A good deal of experience is necessary to recognize these key characters among the great deal of natural variability in most species and this causes much of the mysticism of identification.

Once a conclusion has been drawn, the unknown culture should be compared to one of known identity and with the original species description in the literature.

Computerized keys have been developed recently for several important groups of fungi. These keys can save a lot of time in comparison to going through printed dichotomous or synoptic keys, but they do not help to recognize the correct cultural and micromorphological details required for identification.

The situation is a bit more friendly in the yeasts, especially the yeasts with ascomycetous affinities. As has already been stated above, the ability to grow on a defined set of carbon and nitrogen sources under standardized conditions can be used to identify species.

A complete monograph of the yeasts and a computerized key based on it has been published by Barnett *et al.* (1990).

Commercial identification systems for yeasts are mostly limited to species of clinical relevance and will give wrong results with most other material. They may, however, be used to replace time-consuming growth tests in tubes if it has been assured that both systems will give the same result.

In some groups of fungi as in basidiomycetes, where no identification of cultures is possible, only indirect methods can help establish the identity of the culture and the species in question. Fruiting experiments may lead to carpophores resembling the natural ones. Several cultures isolated in parallel should have a similar appearance, growth rate, pigmentation, etc., but it should always be considered that single spore cultures may differ considerably in geno–and phenotype and can lead to erroneous results.

Molecular methods may also be used with success, e.g., homology studies between DNA from culture and fruit body, restriction analysis of mitochondrial DNA, and enzyme patterns. In pathogenic or symbiotic species, successful colonization of the host may be taken as proof of identity.

PRESERVATION

Considering the amount of time and money spent on the isolation and identification of a culture, it indeed represents not only a high scientific but also a monetary value which should be preserved unchanged in all its properties for a long time. Various techniques have been developed for this purpose, which differ in their technical complexity and their usefulness for different groups of organisms (Smith and Onions, 1983).

The simplest, though not really a preservation, technique is serial transfer onto fresh medium. If not carried out under expert supervision, this highly error-prone method may rapidly lead to irreversible loss of properties like sexuality, pathogenicity, sporulation, etc. By increasing the shelf-life of agar cultures, i.e., retarding the drying of agar by covering with oil or water, the interval between transfer can be extended to about 1 or 2 years.

More sophisticated preservation methods for sporulating fungi and many yeasts are based upon drying of the whole culture or parts of it. This can be done by spray-drying (baker's yeast), drying on silica gel or porcelain beads (yeasts, conidial fungi), or under vacuum with or without prior freezing (1-drying, freeze-drying). Non-sporulating mycelial fungi (comycetes, many basidiomycetes) can only be preserved by freezing. Experience has shown that the lower the storage temperature, the better the survival rate, i.e., best results have been obtained by preservation in liquid nitrogen containers at temperatures of-160°C to -196°C. By miniaturizing the storage technique, a more economical utilization of the liquid nitrogen container can be achieved. A method developed at the DSM makes use of small PVC tubes with a volume of only 50 1 (Hoffmann, 1991).

REFERENCES

Ainsworth, G.C., Sparrow, F.K. and Sussmann, A.S. (eds). 1973. ***The Fungi. An Advd. Treatise.*** New York, USA.

Barnett, J.A., Payne, R.W. and Yarrow, D. 1990. ***Yeasts: Characteristics and Identification.*** Cambridge, England.

Booth, C. (ed). 1971. ***Methods in Microbiology.*** London, New York. Vol. 4.

Gams, W., van der Aa, H.A., van der Plaats-Niterink, Samson, R.A. and Stalpers, J.A. 1987. CBS *Course of Mycology*. Baarn.

Hawksworth, D.L. 1988. Identification. In D.L. Hawksworth and B.E. Kirsop (eds), *Filamentous Fungi. Living Resources for Biotechnology*. Cambridge, England. pp. 100-14.

Hoffmann, P. 1991. *Cryopreservation of Fungi*. WFCC Technical Information Sheet No. 5 *World J. Microbiol. Biotechnol.*, 7, 92-94.

Smith, D. and Onions, A.H.S. 1983. *The Preservation of Maintenance of Living Fungi*. Kew, England.

von Arx, J.A. 1981. *The Genera of Fungi Sporulating in Pure Culture*. Vaduz, Liechenstein.

17

HOW TO GAIN MORE INFORMATION ON THE INTER-RELATIONSHIP BETWEEN CELLS AND CELL ENVIRONMENT IN BIOREACTORS

K. SCHUGERL, K.H. BELLGARDT, A. LUBBERT AND T. SCHEPER

Institute for Technical Chemistry, University of Hanover, Germany

Abstract. In bioreactors, complex inter-relationships exist between the cells and their micro- and macro-environment. The present paper deliberates on modern techniques for the determination of the concentration and biological state of the cells, the evaluation of the micro-environment of immobilized cells and cell aggregates, and the determination of the macro-environment of suspended cells. The usefulness of these techniques is demonstrated on some examples.

INTRODUCTION

In order to achieve the full biological potential of the cells, the optimal environmental conditions of cell growth and product formation must be maintained in the bioreactor, at least with regard to the most important parameters. The prerequisite for this control is the observation of the physical and chemical cell environment as well as the biological behaviour of the cells.

The observation/measuring techniques depend on the type of bioreactor system. In the case of biofilms, cell aggregates and immobilized cells, the local concentration gradients are of importance. Mathematical models with distributed parameters are required for describing these systems. Suspended cells pass different regions of the reactor rapidly, thus, with few exceptions, global reactor properties and mathematical models with lumped parameters usually are sufficient for describing these systems. However, for reactor optimization, specific local information is imperative.

In order to relate the cell behaviour to its environment, information on the concentration and biological state of the cells is also required.

If direct measurement of the key parameter is not possible, it is calculated by means of a model and monitored process variable.

In the following, measuring techniques are considered for the evaluation of this information.

DETERMINATION OF THE CONCENTRATION AND BIOLOGICAL STATE OF THE CELLS

Light Attenuation/Reflexion Techniques for Cell Concentration Monitoring

Cell concentration usually is determined by off-line counting in low-density tissue cultures and gravimetrically in high-density cultures of micro-organisms. The traditional technique for cell mass concentration estimation is the off-line measuring of the optical density of the cell suspension. This technique was adapted for on-line use by measuring the light attenuation in the broth. A simpler type of this equipment compares the intensities of non-attenuated light beam and the light beam attenuated by scattering on the cells. In this way the intensity change of the light source can be compensated for Model 210-F1/130 of Monitek). Another type uses reflected light (e.g., Model ASD 81/AF 44 of Aquasant).

Since the transparencies of the light source and light detector windows change during cell cultivation due to cell growth on the window surfaces, a twin-source/twin-detector system of a different light path length was developed. by alternatively measuring the light attenuation with the four possible source-detector combinations, the drift caused by the change of the window transparencies can be compensated (e.g., Models Mex 3/rd-10/5 of BTG-Anlagetechnik and Model 350/136 of Monitek).

New Techniques for Cell Concentration Monitoring

Several other methods are recommended in the literature, though only a few of them can be used in real cultivation media.

Acoustic resonance densitometry uses the direct relationship between the resonant frequency of the liquid in a fixed volume and its specific gravity (e.g., Model DMA35 of Paar). The same principle is used for measuring the cell mass-concentration in cultivation media. By determining the specific gravity of the cell-containing medium and the cell-free medium, it is possible to determine the cell concentration, provided that no suspended particles or bubbles are in the medium. This technique, however, is unsuitable for on-line measurements. The equipment will be commercialized in the future.

The radio-frequency dielectric properties of microbial suspensions are a direct function of the size and concentration of the cells. Measurement of these properties permits the on-line estimation of the cell concentration in the medium, if no bubbles are present. Equipment is already on the market (Model Bugmeter of Aber Instruments). However, this instrument is very inaccurate at low cell concentrations and cannot be used for *in situ* measurements at high cell concentrations because of the high ion strength of the culture medium.

Culture Fluorescence

NADH and NADPH in the cells can be excited by UV light, and the induced fluorescence can be measured by suitable detectors at 460 nm. This NADH fluorescence has a large potential as a tool for monitoring, studying, and controlling cultivations. During balanced growth, the cell concentration can be determined by means of the NADH fluorescence of the cell on-line. In general a non-linear relationship exists between fluorescence intensity and cell concentration. In special cases (e.g., at unlimited growth of *Zymomonas mobilis*), the relationship between them is linear. If the evaluation of this relationship is performed under similar operation conditions, and the growth is balanced, the agreement between measured and calculated cell concentrations is satisfactory. Two microfluorometers are on the market: Fluoromeasure of BioChem, USA, and [illegible] sensor of Ingold, Switzerland. Both of them have been used to characterize cell mass concentrations and biological activities of the cells.

$NAD^+/NADH$ is a key component in cells. Their ratio changes quickly in response to the cell environment. For example, glucose repression, oxygen limitation, and variation of the cell growth (dilution) rate influence this ratio. Since only the NADH species is fluorophore, it can easily be detected by means of the culture fluorescence, if the fluorescence of the medium is low. This technique is potentially very useful. However, it is necessary to adapt the instrument to the particular medium being used and to carry out calibration. In the presence of fluorescent-medium components or products, it is essential to determine the fluorescence in the call-free medium and in the suspension separately. The combination of these data allows the on-line determination of the cell mass concentration and the variation of oxidative to oxidative-fermentative metabolism of the cells (e.g., *Saccharomyces cerevisiae*) and *vice versa.*

Calorimetric Methods

The heat generated during the cultivation of cells can also be used to determine the growth rate. Since the relationship between the rates of growth, oxygen consumption, and heat generation is linear, it is possible to evaluate the growth rate in batch reactors from calorimetric measurements. One calorimeter is on the market (Reaction Calorimeter RCI of Mettler Instruments) which can also be used for cell concentration measurements.

Determination of Intracellular Enzyme Activity

At times it is important to determine the intracellular enzyme activity on line. Two techniques are known: one of them uses an air-segmented flow system, where the cell membrane is destroyed enzymatically, a model substrate permeates into the cell, converts into a coloured product; the latter permeates out of the cell, is separated by dialysis and detected by a photometer about 20 minutes after the cell suspension is sampled. According to the second technique, the cell is disintegrated by ultrasound, a model substrate is converted into a coloured product, the cell fragment suspension is diluted, and the product concentration is measured in an FIA system by a photometer.

Flow Cytometry

Flow cytometry has been established for more than 15 years as a research tool for eukaryotic cell research, mainly in the field of cell cycle studies and for the off-line determination of the cell size and intracellular substances (DNA, RNA, proteins, lipids). Recently it has been used in biotechnology for the characterization of cell properties. By measuring the length of the aligned cells, it is possible to determine the fraction of the plasmid-free cells, if the plasmid contains an antibiotic resistance gene, and if the cells are cultivated in the presence of antibiotics.

In the case of cells with a temperature-sensitive promotor/ repressor system, the gene expression is induced by a temperature increase from 30 to 42°C. During gene expression, the cell size and intracellular protein content changes, which can be measured with a laser flow cytometer. This allows the quasi-on-line monitoring of gene expression.

In our laboratory, we succeeded in establishing a laser-flow-cytometer system for the on-line determination of size distribution of mammalian cells and cell fragements.

A combination of culture fluorescence and flow-cytometer measurements allows important information on population dynamics of synchronized continuous cell cultures to be gained.

Application of Software Sensors and Observer/Filter

The simplest application of software sensors is the use of material balance of O_2 and CO_2, which can be monitored to calculate, for example, the volumetric mass transfer coefficient, k_La or respiratory quotient, RQ.

If we want to calculate, for example, the cell concentration of the monitored medium components in synthetic medium, we can use a simple relationship between substrate and oxygen consumption and CO_2 production rates. For complex cultivation medium, we need a deterministic (observer) or stochastic (filter) model.

If we want to gain more information on the regulation of the cell, we need structured mathematical models and must identify the model parameters based on the monitored data. The models can also be used for process control. The calculated data, however, must be compared to those measured. In order to avoid deviation between them, the model parameters of the state estimator must be frequently re-identified (adaptive control).

CHARACTERIZATION AND MODELLILNG OF MICRO-AND MACRO-ENVIRONMENT OF THE CELLS

Micro-Environment and State of the Cells

Molds often form pellets, which can be of various sizes and densities. The process performance depends considerably on the pellet properties. If the pellets are small enough, the entire pellet consists of metabolically active cells, because the cells can be supplied by substrate and oxygen.

In the centre of large pellets, the cells die and are autolyzed owing to the lack of substrate and/or oxygen. The pH, substrate, oxygen, and cell concentration profiles were measured to evaluate the micro-environment, and the cytoplasm of the cells were stained to determine the biological state of the cells in different positions in the pellet. This allowed the development of a pellet model for optimizing the production process.

Similar measurements were also carried out in immobilized cell systems. In alginate pellets, the local measurements are much more difficult to perform. The viability of the immobilized cells on the pellet surface, however, can be measured *in situ* by NADH fluorescence, as Reardon, Scheper and Bailey were the first to find out.

Chemical Macro-Environment of the Cells: Monitoring Of Medium Composition

Except for the pH value, dissolved oxygen partial pressure P_{O2}, and sometimes the disolved CO_2 partial pressure, P_{CO2}, which are measured by *in situ* sensors, medium components are determined outside the reactor by means of an on-line aseptic sampling system and common analytical instruments.

On -line aseptic sampling is now a well-established technique. Two systems are on the market: BIOPEM of Braun and Diessel, Melsungen, which is a stirred cross-flow filtration module integrated into a medium recirculation loop, and the *in situ* FILTER of ABC

Bioverfahrenstechnik GmbH, Puchheim; both have been used for several years in industrial practice.

Several other systems have been discussed in publications, some of them are offered on the market, it seems without much success.

Of the classical analytical techniques, four have been applied successfully in biotechnology: continuous-flow air-segmented analyser, CPA; flow injection analyser, FIA; HPLC; and mass spectrometry, MS.

Continuous-flow air-segmented analysis (CFA) is popular in the chemical industry (the Technicon system of Bran + Lubbe, Norderstedt and the CFA system of Skalar Analytika, Breda). This technique, however, if used for cultivation-medium sampling, usually breaks down after three days of operation owing to protein precipitation and cell growth in the analyser system; regular rinsing of the system with bactericide solution permits a long-range operation, yet the system is quite susceptible to breakdown. Other shortcomings are the high amounts of reagents and samples and the long analysis time this technique requires.

CFA is also used in monitoring extracellular enzyme activities with model substrates, which are converted into coloured products and detected by a photometer.

Flow injection analysis, FIA, uses a carrier flow with a reagent into which small amounts of sample pulse is injected. This causes a large dilution of the cultivation medium by which the probability for cell growth in the analyser system is reduced considerably.

The analyte forms a product with the reagent, its concentration peak height is measured in the detector. Requiring low amounts of reagent and sample, short reaction times, and high flexibility are properties leading to its increasing popularity.

Immunoassays were developed and integrated into the FIA system to measure the concentration of biologically active proteins.

They have also recently been used for on-line monitoring of the protein concentration in cultivation media. Three types of immunoassay FIA systems are employed for this purpose: turbidometric, homogeneous competitive, and heterogeneous competitive immunoassays.

In the turbidometric immunoassay, the antigen (Ag) in the sample is mixed with the reagent (antibody Ab), and the concentration of the antigen-antibody (Ag-Ab) complex, which forms aggregates and causes light extinction, is measured by an optical detector. This is a fast technique (3-5 minutes response time), but with restricted sensitivity.

In the homogeneous competitive immunoassay, the Ag in the sample competes with the TRITC (tetramethy l- rhodanine-isothiocyanate)- labelled Ag in reagent 1 for the binding sites of the FITC (fluorescein-isothiocyanate)- labelled Ab in reagent II. The concentration of the labelled Ag-labelled-Ag complex is measured in a fluorescence detector by means of its fluorescence emission caused by the energy transfer between labelled-Ag and labelled-Ab in the complex. In the unlabelled-Ag-labelled-Ab, no energy transfer is possible, therefore no signal appears in the fluorescence detector. The higher the fluorescence signal, the lower the concentration of Ag in the sample.

In the heterogeneous competitive immunoassay, Ab is immobilized in a cartridge. The unlabelled Ag in the sample and the FITC-labelled Ag in the reagent compete for the binding sites of Ab. They form an Ab-Ag- complex in the cartridge. After rinsing the cartridge, the unlabelled Ag and labelled Ag are displaced, and the concentration of the labelled Ag is measured in a fluorescence detector. Again, the higher the signal, the lower the Ag concentration in the sample.

High Performance Liquid Chromatography (HPLC) is a standard technique in the chemical industry. It is, however, rarely applied in online monitoring in biotechnology. When using very pure chemicals and degassed solvents, stable, long-range operation is possible. The advantage of this technique is that several medium components can be monitored by one single item of equipment.

HPLC is also employed in the analysis of proteins. Recently, superfast HPLC techniques were developed for protein analysis, mainly in the Horvath Laboratory at Yale University. This technique, however, has not been used yet for on-line monitoring of proteins.

Inexpensive mass spectrometers (MS) have become popular in monitoring the exhaust gas composition of cultivations. Dissolved gas and volatile components have also been analyzed on-line with this technique by using on-line pervaporation sampling.

Mass spectrometric analysis of bioploymers was not possible until recently because of the high polymer masses. Different ionization techniques have been developed of late, that extend the used of the mass spectrometer to bioploymers. Polymers with several charges can be analysed in the MS. The main problem is the production of ions with definite charges and the identification of molecules based on their charge to the mass ratio.

By fast atom bombardment (FABMS), biopolymer ions with neutral atoms are produced. A useful range is up to 10 kDa.

Liquid secondary ion mass spectrometry (LSIMS) uses energetic ions (e.g., CS^+) for ionization.

Electrospray mass spectrometry (ESMS) employs a high voltage field to produce a dispersion of charged fluid droplets. Multicharge protein ions desorb from the fluid surface by gas (ionspray) or by heat (thermospray) and can be detected.

So far, on-line monitoring of bioploymers by mass spectrometry has not been performed.

These ionization techniques will extend the application of MS for biopolymers in the future.

Physical Macro-Environment of the Cells: Flow Structure and Transport Processes

On account of the low solubility of oxygen in the culture medium, a permanent transfer of oxygen into the medium is necessary to supply the aerobic cells with oxygen. The oxygen is supplied by dispersion of the air into the cultivation medium.

This is done by dissipating the energy in the liquid medium, and controlled by the amount of specific power input into the reactor.

In order to supply the micro-organisms with an optimal environment for cell growth and product formation in the entire reactor, diverse transport processes are used at different scales. An abundant circulation flow is induced into bioreactors for a fast but coarse distribution of substrates including oxygen, fed into the bioreactor at one particular location. Turbulence is required to support thorough mixing and to promote the mass transport in the vicinity of the cells. The integral transport is characterized by residence time or circulation time measurements, thorough mixing, for example, by local stimulus response techniques.

All techniques used were developed in terms of model-supported measuring techniques.

For the evaluation of the global properties of the phases, stimulus-response techniques were used in the entire reactor, which yielded the distribution of residence times in the gas phase as well as the liquid and gas circulation times.

The characterization of the local gas phase behaviour is performed by an ultrasound Doppler impulse anemometer, which provides the local gas content, bubble velocity vector, and gas/liquid specific interfacial area; all of these were measured *in situ* during cultivation.

The pseudostochastic heat-pulse technique is used for the *in situ* characterization of the local liquid phase behaviour in the reactor during the cultivation. This technique provides the local liquid velocity vector and mixing.

Both techniques have been used to evaluate the spatial distribution of these vectors (gas and liquid flow pattern) in stirred tanks and airlift tower loop reactors.

These measurements render important information on the fluid-dynamic performance of bioreactors and its dependence on the reactor construction, stirrer type, and speed as well as aeration rate.

EXAMPLES OF THE APPLICATION OF THE INFORMATION GAINED BY *IN SITU*, ON-LINE, AND OFF-LINE MEASUREMENTS

Monitoring of Cell Behaviour by a Microfluorometer

During baker's yeast production in a fed batch mode, the culture fluorescence combined with the CO_2 production rate has been used by Meyer and Beyeler to control the sugar-feeding rate, to avoid the change of the oxidative into an oxidative-fermentative metabolism..

Control of Micro-Environment by Evaluation of Pellet Properties

In order to reduce the viscosity of the *Penicillium chrysogenum* mold, the production was performed by pellet morphology. However, with pellets of 1 mm diameter, the productivity was low.

Through histological investigations and by measuring the dissolved oxygen concentration distributions in the pellets, it was possible to optimize the pellet size. With optimized pellets, the productivity was increased by a factor of two, and the specific power input was decreased by a factor of three.

Control of Chemical Macro-Environment: Regulation of Biosynthesis by Parameter-Adaptive Control of Dissolved Oxygen Concentration

During the production of Cephalosporin C by *Cephalosporium acremonium* (CPC), the concentration of CPC was reduced by a factor of seven and the concentration of penicillin N (PEN N) was decreased by a factor of three, provided the dissolved oxygen concentration was reduced from 40% to 5% of its saturation value.

Control of Chemical Macro-Environment: Regulation of Biosynthesis by Control of Medium Composition

After the growth phase, the CPC formation was initiated by reducing the sugar consumption rate below a critical value. The biosynthesis stopped when the sugar consumption rate was increased above this critical value, and could be restarted by decreasing the sugar concentration below it.

Influence of Flow Pattern on Biosynthesis

In the presence of 100 g/1 peanut flour in *Cephalosporium acremonium* cultivation medium, the viscosity increases during the growth phase. Above a critical viscosity, the medium recirculation rate stops in an airlift tower loop reactor. This causes the reduction of the dissolved oxygen concentration to zero. As a result, the ring expandase is repressed, the CPC concentration is reduced, and the PEN N concentration is increased. After the morphology change from the long hyphae to arthrospores, the viscosity diminishes, the cultivation recirculation restarts, the dissolved oxygen concentration increases, the production rate of the CPC increases and that of PEN N diminishes.

CONCLUSION

The measuring techniques in biotechnological practice have been improved considerably in recent years. On-line process analysis has been extended to several key components of the cultivation media. Intracellular components can be analyzed, and the concentration of otherwise unmeasurable components can be evaluated by monitoring other components. This development has enhanced our information on cell-bioreactor systems and also augments our knowledge on the behaviour of the cells under industrial production conditions as a function of their micro- and macro-environments.

REFERENCES

Several papers deal with particular techniques of this topic. Rather than looking for the original literature, one can use surveys on the entire subject. The most up-to-date information on this topic is given at present in the book:

1991. Measuring, Modelling and Control in Biotechnology. In H.J. Rehm and G. Reed (eds) (vol ed K. Schugerl), *Biotechnology*, VCH, Weinheim. 2nd edn, Vol. 4.

18

BIOREACTOR EVALUATION

I. ADLER
Henkel KGaA, Dusseldorf, Germany

INTRODUCTION

Biotechnological products are found in widely differing markets. Ranked according to turnover, waste water treatment and alcoholic beverages hold first place, each holding more than 35% of the world turnover (1980). They are followed by pharmaceutical products with 21.4%. The remaining 7.3% are covered by food and feed (except cheese and bread), fine chemicals and enzymes. The demands on biotechnological equipment especially bioreactors therefore have to cover the following three areas:

1. construction of very large bioreactors for non-sterile operation for waste water treatment
2. bioreactors for non-sterile but almost germ-free operation and high demands on the material typical for food industry
3. bioreactors for sterile operation and with high instrumentation for microbial processes with classically or genetically recombined organisms as well as for cell cultures.

In parallel with this broad spectrum of biotechnological processes, the existence of products with low added-value (e.g., waste water treatment, leaching, etc.), and with high added value (e.g., diagnostic and therapeutic products) has to be recognized. It has an important influence on the economy of scale (Table 18.1). Besides quality of material and sterility, added-value, and scale, the kinetics of the biological process determine the demands on an optimal bioreactor. Praxis teaches that even nowadays the technical use of the whole kinetic potential of a cell leads to difficulties. Often, the kinetics of the coupled reactions are not completely known or else knowledge of the parameters determining velocity or regulation is missing. Both have to be known in order to design the optimal environment for the cell.

TABLE: 18.1
FIELDS OF APPLICATION AND TOTAL VOLUME OF BIOREACTORS.

Field of apllicn./product	*Reactor volume (m^3)*
* Waste water treatment	some 1000 to some 10000 (e.g., HOECHST Bio Hochreaktor: 18000)
Single cell protein (SCP)	(e.g., ICI: 2300)
*Acetic acid, fodder yeast, ethanol, citric and lactic acid, baker's yeast, penicillin	some 10 to 500
* Laboratory/pilot plant	20 1 to some m^3
* Laboratory	up to some 1

HOW TO QUALIFY THE PERFORMANCE OF BIOREACTORS

The aim of all biotechnological processes is a technical use of the biosynthetic potential of the cell as complete as possible. The cost of reaching this aim can vary widely. Often there are apparent limits. In order to discover these limits, the diversity of evaluating and influencing parameters are arranged in a matrix with the dimensions 'construction and operation parameters' (Fig. 18.1). The elements of this matrix have to be determined from experiments with biological test systems and form the basis for evaluation of a bioreactor.

The prerequisite for the evaluation of a bioreactor is the knowledge of the parameters that influence growth and product formation of the microbial cell. In a process development for instance experiments to heighten product concentration by enhancement of substrate, concentration by a factor of 5 or 10 may lead to problems with growth and/or product formation, although the medium on the low concentration level has been optimized. These problems may arise from components that are now on an inhibiting concentration level, from trace elements that now have a limiting concentration, from a change in regulation or from a too high osmotic pressure. Often, not all parameters that influence growth and product formation are known. Besides intracellular biological parameters describing the potential of the cell, there are a number of other influencing parameters that can be divided into the categories 'chemical' and 'physical parameters' (Fiechter, 1984). It is, therefore, recommended that the known effects be sorted into the categories shown in Table 18.2 and the experiments run so that the effects and the causes from which they derive can be determined.

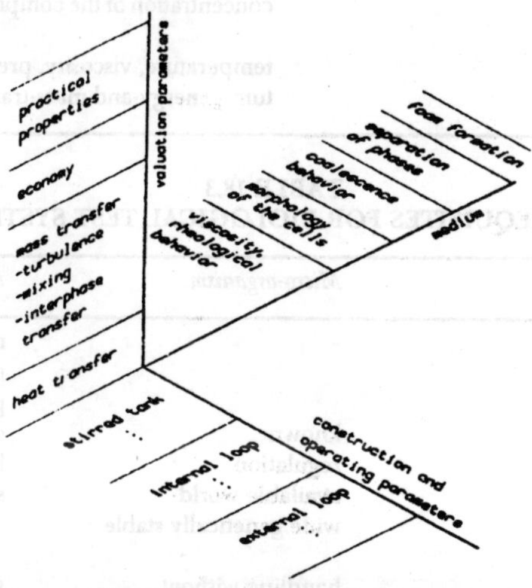

Fig. 18.1: To sort the diversity of evaluating and influencing parameters they are arranged in a matrix with the dimensions 'construction and operating parameters', 'medium properties' and 'valuation parameters'. The elements of this matrix have to be determined from experiments with biological test systems and from the basis for evaluation of a bioreactor.

A biological test system used for comparing bioreactors has to fulfil several basic prerequisites which can be divided into four areas (Table 18.3):

1. Transferability of results
2. Reliability
3. Reproducibility
4. Practicability.

TABLE: 18.2
PARAMETERS AFFECTING GROWTH AND PRODUCTION OF MICRO-ORGANISMS

Intracellular biological	**Genetic information (potential of the cell) regulation of transcription and translation (induction, repression) metabolic regulation (activation/inhibition by products and metabolites)**
Extracellular chemical	**components (C- & energy source, nutrient salts, trace elements, growth factors, products) concentration of the components pH-value**
physical	**temperature, viscosity, pressure, momentum-, energy-and mass-transfer**

TABLE 18.3
PREREQUISITES FOR BIOLOGICAL TEST SYSTEMS

Prerequisite	*Micro-organism*	*Medium*
Transferability of results	–	representative physical properties
Reliability	known regulation	clear limitation
Reproducibility	available world-wide genetically stable	synthetic
Practicability	handling without problems, lowest classification of hazard	simple and cheap

The comparison experiments have to be run under conditions that are relevant to the technical production process of interest. The transferability of the experimental data from laboratory or pilot scale up to production scale has to be warranted. This means that medium properties especially coalescence behaviour, rheological parameters, and morphology of the cells, have to be representative of the practical process. The experimental conditions must allow the major valuation parameters to be checked.

The performance of the process which is given, for example, by maximum product concentration, substrate yield, or productivity, can be influenced by effects from three different parameters:

1. by apparative parameters, i.e., the bioreactor: mass transfer, mixing, and shear rate
2. by chemical parameters, e.g., hidden limitation
3. by biological parameters, e.g., change of metabolism, change of coalescence behaviour by ethanol formation.

A clear separation of these effects is the prerequisite for reliable judgement. The biological parameters as well as regulatory effects that arise from inhibition or limitation have to be known. All growth limiting components of the medium have to be fixed and the excess of the other components has to be evident.

The experimental data have to be reproducible in order to check and extend them at other places and in other plants. Genetic stability of the strain is requested. Changes in genetic potential during the experiments and during long-term preservation in stock cultures must be avoided.

It must also be possible to prepare the medium at different places with the same quality. Therefore, completely synthetic media that also have advantages for the control of metabolic effects are preferred.

Practicability of the test systems means handling of the cells without problems and a strain of the lowest biosafety class. The medium must be simple to prepare and cheap. The specific turnover rates and the substrate yield should be large to allow cheap experiments in a limited time.

This shows that biological test systems have to fulfil rigorous prerequisites. Actual production systems are most likely to be applicable only in individual cases.

METHOD FOR TESTING

In order to determine the expenses necessary for the technical use of the biosynthetic potential of the cell and the possible limits that may arise, biological test systems have been developed. They consist of the yeast *Trichosporon cutaneum*, the fungus *Chaetomium cellulolyticum* and the bacterium *Escherichia coli B/r*.

The medium properties of the yeast system can be varied in a wide range by addition of chemical antifoam agent and xanthan as thickener. Foaming as well as non-foaming low-viscous media and high-viscous media with pseudoplastic behaviour are possible. Thus, the medium properties can be adjusted to be representative of an actual production process (Table 18.4).

Figure 18.2 shows the batch culture of the yeast *Trichosporon cutaneum* in the non-foaming medium in a 0.3 m^3 stirred tank bioreactor. The oxygen transfer rate marked by the arrow is defined as characteristic for the adjusted operating parameters.

By varying gas flow rate and stirrer speed within the determined limits, a group of curves can be derived which represents the oxygen transfer capacity and the biomass productivity for the investigated bioreactor (Fig. 18.3). The composition of the synthetic medium is given in Table 18.5.

TABLE: 18.4
BIOLOGICAL TEST SYSTEMS, VARIATION OF MEDIUM PROPERTIES.

Low viscosity, without foam formation:

1. *Trichosporon cutaneum* with antifoam agent (1-3% polypropylene glycol P 2000)
2. *Escherichia coli* with antifoam agent (1% polypropylene glycol P 2000)

Low viscosity, with foam formation:

3. *Trichosporon cutaneum* without antifoam agent

High viscosity, without foam formation:

4. *Trichosporon cutaneum* with thickener (up to 3% xanthan, pseudoplastic behaviour)

High viscosity, due to cell morphology:

5. *Chaetomium cellulolyticum* (mycelial growth)

TABLE: 18.5

Composition of the synthetic medium for *Trichosporon cutaneum*. Compared to medium D (Hug *et al.*, 1974), the relative amount of salts has been enhanced by 30%. The concentration of glucose as well as all other components has been enhanced several times. This allows an oxygen uptake rate of the culture that is always larger than the achievable oxygen transfer rate. The given amounts are for 1 m^3 of medium.

Component	Amount	Unit	Component	Amount	Unit
Glucose. H_2O	110.00	kg	$FeCl_3.6H_2O$	62.38	g
$(NH_4)_2SO_4$	26.00	kg	$MnSO_4.2H_2O$	45.50	g
$(NH_4)_2HPO_4$	8.33	kg	$ZnSO_4.7H_2O$	39.00	g
KCl	3.78	kg	$CuSO_4.5H_2O$	10.13	g
$MgSO_4.7H_2O$	1.95	kg	m-Inositol	260.00	g
$CaC1_2.2H_2O$	1.23	kg	Ca-panto-thenate	130.00	g
Polypropylene glycol 2000	1-4.5	l	Vitamin B1	26.00	g
			Vitamin B6	6.5	g
			Biotin	0.13	g

The time course of the culture without antifoam agent yielding a foaming system is only slightly different from that with chemical antifoam agent (Fig. 18.4).

In the presence of xanthan as thickener, oxygen transfer is reduced significantly. This allows the batch culture to be run with stepwise-adjusted operating parameters and collection of data for different operating parameters, e.g., stirrer speed during one single batch culture (Fig. 18.5).

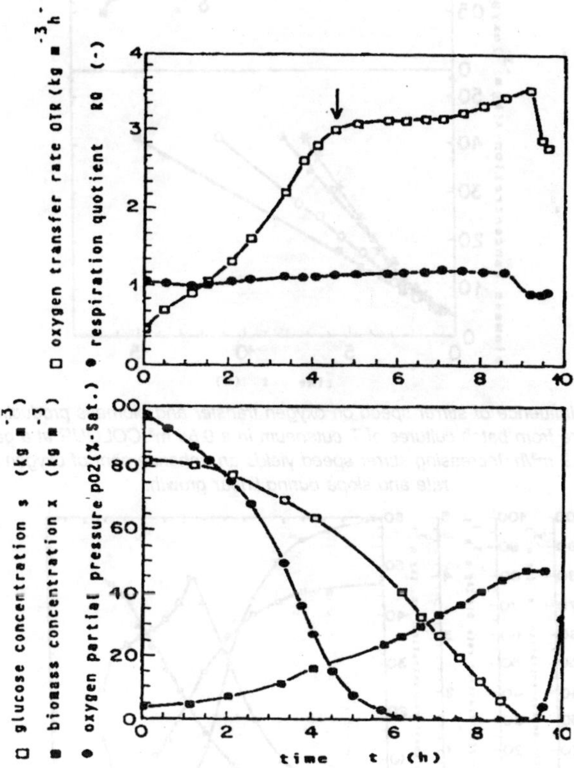

Fig. 18.2: Batch culture of Trichosporon cutaneum in a non-foaming medium in a $0.3m^3$ stirred tank bioreactor (gas flow rate 17.5 m^3/h, stirrer speed: 600/min). Oxygen limitation starts at t=4.5 h. Cell growth continues at initially constant oxygen transfer rate. The oxygen transfer rate marked by the arrow is defined as characteristic for the adjusted operating parameters.

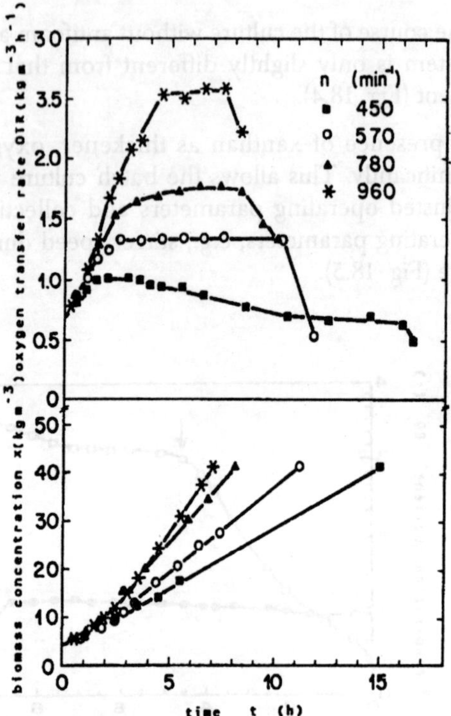

Fig. 18.3: Influence of stirrer speed on oxygen transfer and biomass productivity. The curves are from batch cultures of T. cutaneum in a 0.55 m^3 COLOUR at a gas flow rate of 13.9 m^3/h. Increasing stirrer speed yields an enhancement of oxygen transfer rate and slope during linear growth.

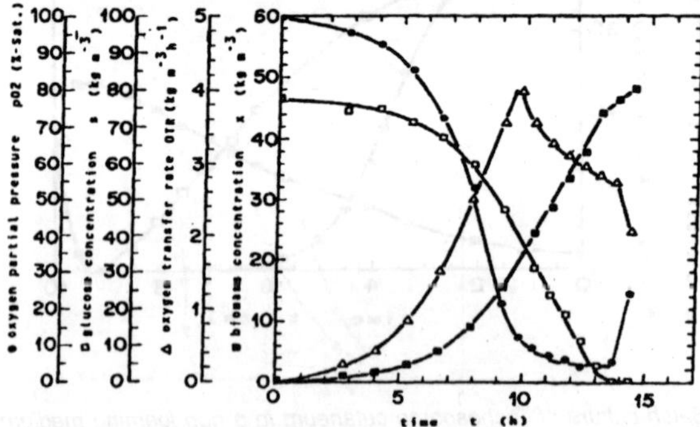

Fig. 18.4: Batch culture of T. Cutaneum without chemical antifoam agent in a horizontal loop reactor (TORUS, liquid volume: 0.065 m^3, gas flow rate 4.2 m^3/h, stirrer speed: 475/min). Beginning oxygen limitation, oxygen transfer rate runs through a maximum defined as characteristic for the adjusted operating parameters. Biological parameters and time course of the other parameters are the same as for the medium with chemical antifoam agent.

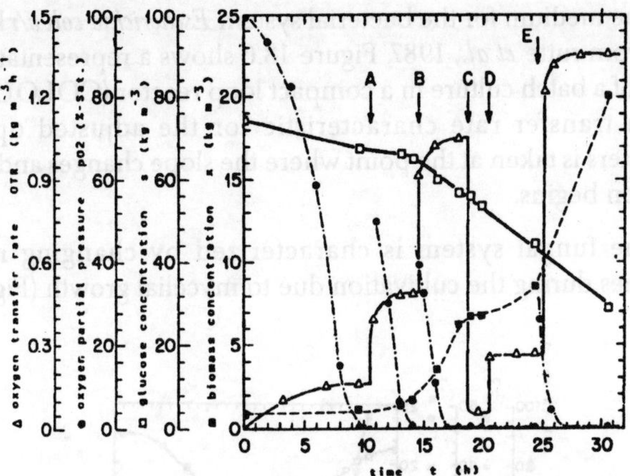

Fig. 18.5. Batch culture of T. cutaneum in a medium with 0.5%w/v xanthan as thickener (horizontal loop reactor, TORUS). Because of reduced oxygen transfer, oxygen limitation occurs rapidly. Therefore, the initial glucose concentration is enough for adjustment of several operating conditions ('step-wise batch'). 2 to 4 h after adjustment the characteristic oxygen transfer rates are detected. Variation of stirrer speed:

A:	200---> 400/min	D: 100--->	300/min
B:	400--> 600/min	E: 300-->	700/min
C:	600--> 100/min		

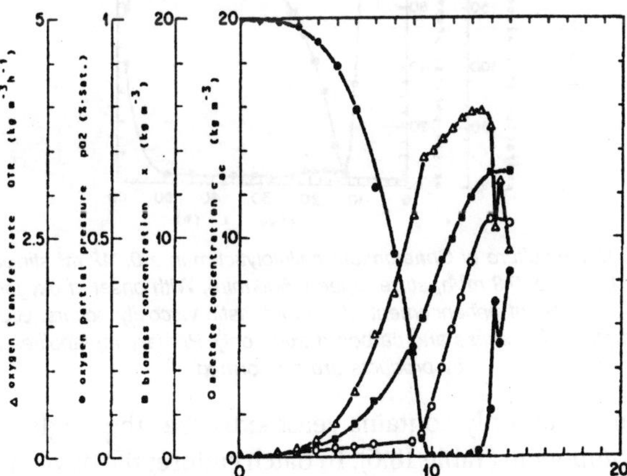

Fig. 18.6: Batch culture of Escherichia coli B/r with chemical antifoam agent. Besides cell growth, acetate formation occurs, especially during oxygen limitation. This leads to a lower biomass yield. The oxygen transfer rate characteristic for the adjusted operating parameters is taken at the point where the slope changes and oxygen limitation begins.

The medium for the bacterial system *Escherichia coli b/r* has been taken from rutte *et al.*, 1987. Figure 18.6 shows a representative time course of a batch culture in a compact loop reactor (COLOUR). The oxygen transfer rate characteristic for the adjusted operating parameters is taken at the point where the slope changes and oxygen limitation begins.

The fungal system is characterized by changing medium properties during the cultivation due to mycelial growth (Fig. 18.7).

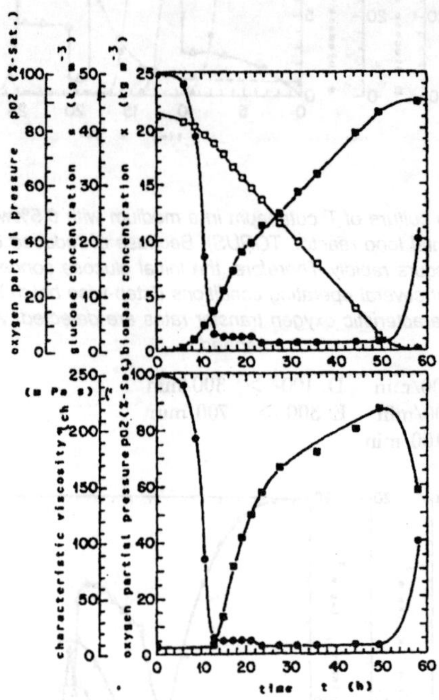

Fig. 18.7: Batch culture of Chaetomium cellulolyticum in a 0.042 m³ stirred tank reactor (gas flow rate: 0.9 m³/h, stirrer speed: 600/min). With onset of oxygen transfer limitation, a significant enhancement of characteristic viscosity occurs. Glucose is used for formation of biomass and carbon dioxide only. Primary metabolites and other by-products aro not formod.

The medium only contains yeast extract as the single complex medium component (Table 18.6). In batch cultures the oxygen transfer rate shows a time course which is typical for the adjusted stirrer speed.

Three regions can be derived where each represents a different morphology of strain due to different mechanical stress for the cell. At low stirrer speeds, pellet formation occurs (region 1).

Moderate stirrer speeds (region II) lead to mycelial growth with long hyphae and low branching. At high stirrer speeds, mycelium with short hyphae and high branching is formed (Fig. 18.8).

TABLE: 18.6
MEDIUM FOR *CHAETOMIUM CELLULOLYTICUM*. CONTAINS POLYPROPYLENE GLYCOL P 2000 AS CHEMICAL ANTIFOAM AGENT. AMOUNTS ARE FOR 1 M³ OF MEDIUM.

Glucose. H_2O	55.0	kg	$FeSO_4.7H_2O$	25.0	g
$(NH_4)_2 SO_4$	15.0	kg	$MnSO_4.H_2O$	8.0	g
KH_2PO_4	10.0	kg	$ZnSO_4.7H_2O$	7.0	g
$CaCl_2.2H_2O$	2.0	kg	$CoCl_2$	10.0	g
$MgSO_4.7H_2O$	1.5	kg	Antifoam	1.0	l
Yeast extract	5.0	kg			

Table 18.7 gives the parameters of the biological test systems as determined from batch cultures.

TABLE 18.7
PARAMETERS OF THE BIOLOGICAL TEST SYSTEMS (DETERMINED FROM BATCH CULTURES).

Micro-organism	μm (/h)	q_{o2},max (mol/kg/h)	$Y_{x/02}$ (kg/kg)	$Y_{x/s}$ (kg/kg)	RQ (mol/mol)
Trichosporon	0.35+0.03	8+1	1.9+0.1	0.52+0.03	1.3+0.1
cutaneum	0.39+0.06	10+1	2.0+0.1	0.55+0.03	1.2+0.1
(low viscosity)	0.35+0.03	6	2.0+0.2	0.53+0.03	1.2+0.1
Escherichia coli B/r	0.53+0.08	23	0.8 (acetate formation)	0.44+0.09	1.1
Chaetomium cellulolyticum	0.29+0.03	8+2	1.6	0.51+0.04	1.1

Comparison of Selected Reactor Types

Stirred tank bioreactors (H/D=3) as well as three loop reactor types have been investigated (Fig. 18.9). This selection was given by the available reactors at the site of investigation (Institute of Biotechnology, Swiss Federal Institute of Technology, Zurich, Switzerland). The total volume of the reactors covers a range from 0.03 to 3.00 m³.

In the order of magnitude from 0.1 to 0.4 m³, a direct comparison of the four reactor types is available.

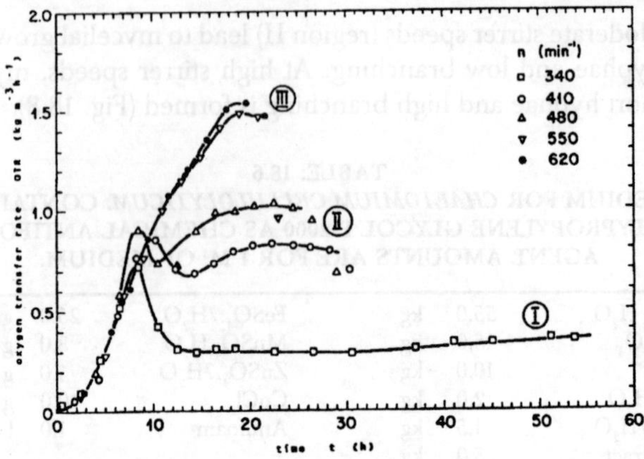

Fig. 18.8: Oxygen transfer rate in batch cultures of Chaetomium cellulolyticum. The time course at different stirrer speeds is characteristic. Region I is characteristic for pellet growth. In the other regions, mycelial growth occurs with long thin hyphae and low branching in region II, and short thick hyphae and higher branching in region III. The data are from a 0.3 m³ stirred tank bioreactor.

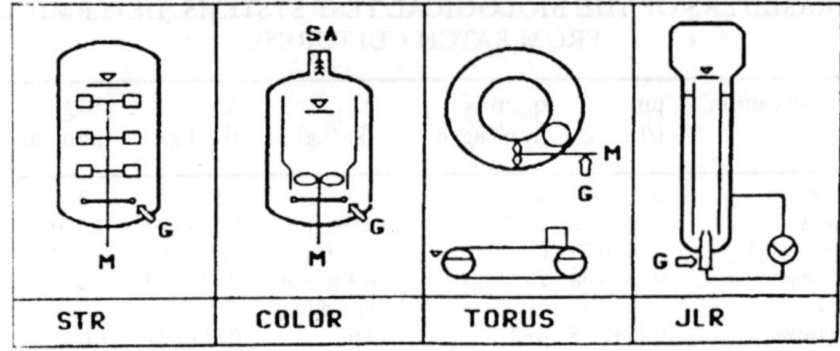

Fig. 18.9: Investigated bioreactor types: stirred tank bioreactor (STR; H/D = 3), compact propeller loop reactor (COLOUR), horizontal loop reactor (TORUS) and jet loop reactor (JLR, H/D = 10). M = motor, SA = mechanical foam breaker, G = gas inlet.

Figure 18.10 shows oxygen transfer performance of the investigated bioreactor types in low viscous media without foam formation and with *Trichosporon cutaneum* as biological system. COLOUR and TORUS yield the highest oxygen transfer rates: 5.2 and 5.6 kg/m³/h, respectively, followed by 4.2 kg/m³/h for the stired tank reactor (gas flow: 2 vvm; overpressure according to pressure drop due to exhaust gas filter). If power input is taken into account, the jet loop reactor gives best results, because high oxygen transfer rates are achieved with low power input. The performance areas of STR and TORUS largely overlap.

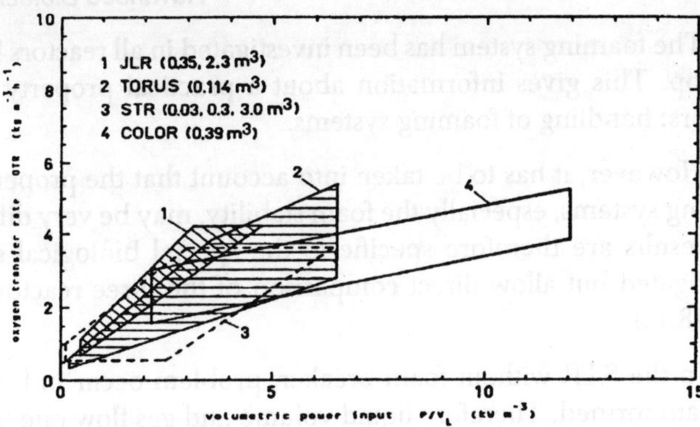

Fig. 18.10: Oxygen transfer capacity of the four investigated bioreactor types in low viscous medium without foam formation. Highest oxygen transfer rates are achieved in the TORUS and COLOUR. The tall construction of the jet loop reactor (JLR) allows the used of hydrostatic pressure and yields high oxygen transfer efficiency. The performance areas of STR and TORUS largely overlap.

The comparison of the results with the yeast and bacterial systems is shown in Fig. 18.11. The change of the biological test system at constant viscosity and coalescence behaviour, i.e., the parameters important for mass transfer held constant, does not change the experimental results. Therefore, for further investigations the application of the yeast system is recommended owing to higher substrate yield and lack of regulatory effects such as acetate formation. The medium without antifoam agent is foaming and yields higher oxygen transfer rates.

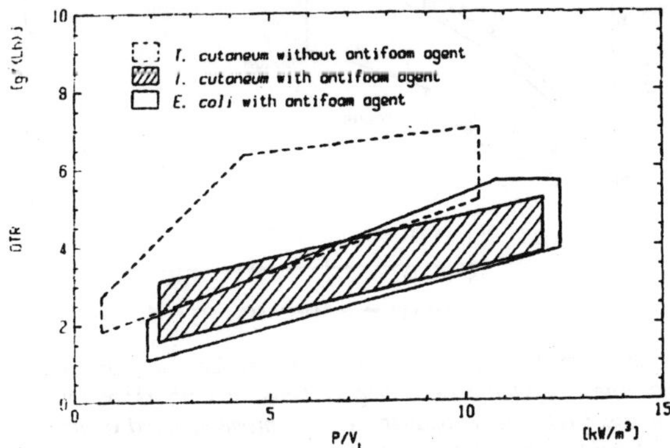

Fig. 18.11: Change of the biological test system for constant viscosity and coalescence behaviour does not change the experimental results. With a foaming system (T. cutaneum without antifoam agent) higher oxygen transfer rates are possible (COLOUR, reactor volume: 0.39 m).

The foaming system has been investigated in all reactors but the jet loop. This gives information about a practical property of the reactors: handling of foaming systems.

However, it has to be taken into account that the properties of foaming systems, especially the foam stability, may be very different. The results are therefore specific to the special biological system investigated but allow direct comparison of the three reactor types (Fig. 18.12).

In the STR without foam breaker, problem occur in handling the foam formed. Therefore liquid volume and gas flow rate have to be reduced significantly. This leads to moderate oxygen transfer at high power input. With mechanical foam breaker, the values are in the same area as those of COLOUR and TORUS (both with mechanical foam breaker) but not as high . For reasons of measuring technique, the power input of the foam breaker is not taken into account.

Addition of xanthan as thickener yields high viscous systems with pseudoplastic behaviour. With increasing amounts of xanthan, decreasing oxygen transfer performance is observed. The results for the TORUS are shown in Fig. 18.13 and compared with those from low viscous systems.

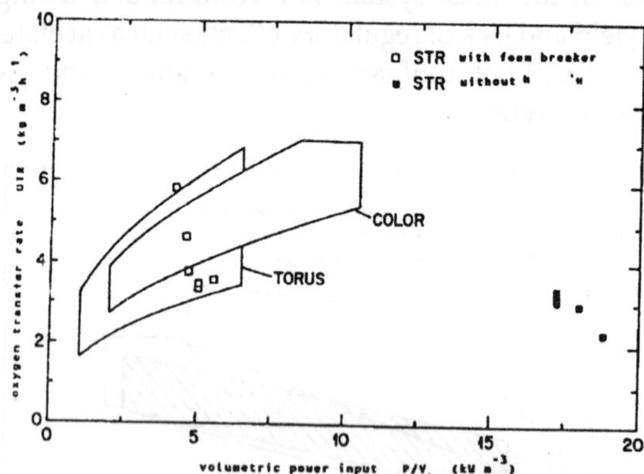

Fig. 18.12: Foaming systems prevent bubble coalescence and therefore allow higher oxygen transfer rates than the non-foaming systems, as long as the formed foam can be disturbed. In the STR without mechanical foam breaker, liquid volume and gas flow rate have to be reduced significantly. This leads to moderate oxygen transfer at high power input. With mechanical foam breaker, the values lie in the same area as those of COLOUR and TORUS (both with mechanical foam breaker) but are not as high. For reasons of measuring technique, the power input of the foam breaker is not taken into account.

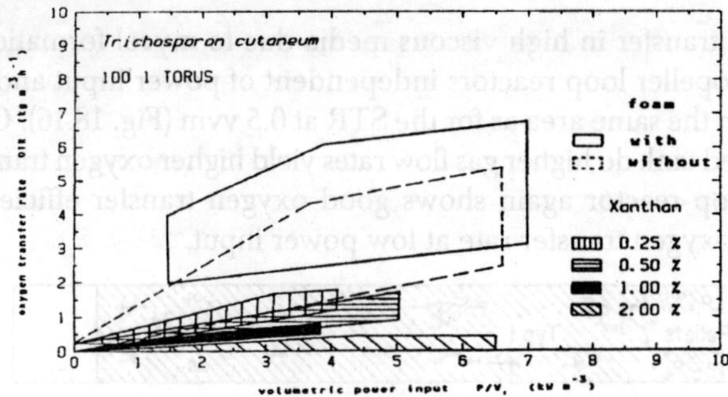

Fig. 18.13: Oxygen transfer performance of the TORUS with low and high viscous media. Increased xanthan concentration diminishes oxygen transfer rate. The influence of power input becomes less important.

Figure 18.14 shows the oxygen transfer capacity of COLOUR and TORUS with media containing 0.5 and 2.0%w/v xanthan. The performance regions change their location according to increased viscosity. The difference in the maximum values is not significant, being due to measuring errors.

The mycelial system changes its morphology with changing mechanical stress. Figure 18.15 shows that a qualitative correlation with stirrer tip velocity as a measure for shear stress is possible. The jet loop reactor is marked at a position corresponding to the tip velocity of the pump-wheel. As only mycelium of type 2 is observed, liquid jet velocity at the inlet obviously is more important as it has a magnitude of about 30 m/s.

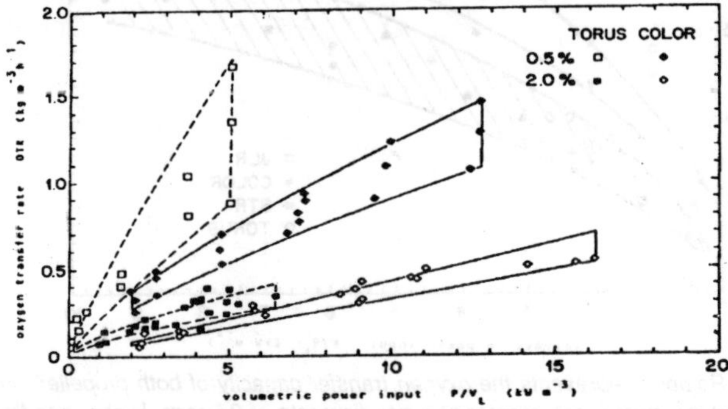

Fig. 18.14: The performance regions for oxygen transfer in COLOUR and TORUS at 0.5 and 2.0% xanthan in the yeast system change their location according to increasing viscosity. Also in high viscous systems, COLOUR needs higher power input than TORUS for the same oxygen transfer rate. The difference in the maximum values is not significant being due to measuring errors.

Mass transfer in high viscous media due to mycel formation is in both propeller loop reactors independent of power input and gas flow rate in the same area as for the STR at 0.5 vvm (Fig. 18.16). Only in the stirred tank do higher gas flow rates yield higher oxygen transfer. The jet loop reactor again shows good oxygen transfer efficiency, i.e., good oxygen transfer rate at low power input.

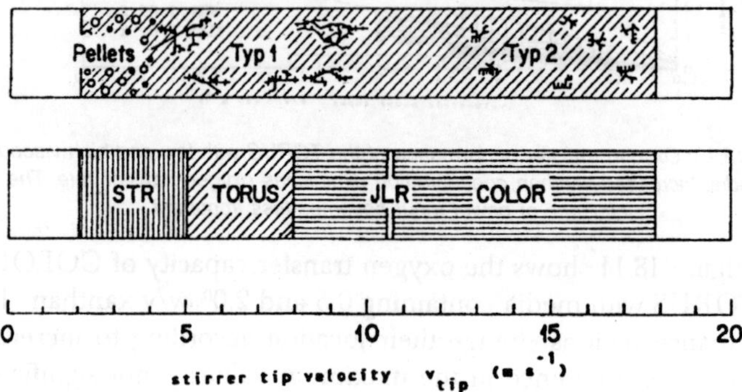

Fig. 18.15: Morphology of Chaetomium cellulolyticum correlates well with stirrer tip velocity as a measure for shear rate. Since in the jet loop reactor (JLR) only mycelium of type 2 is observed, the tip velocity of the pump wheel as indicated is not important but the relative velocity between liquid jet and medium, 32.4 m/s, is significant instead.

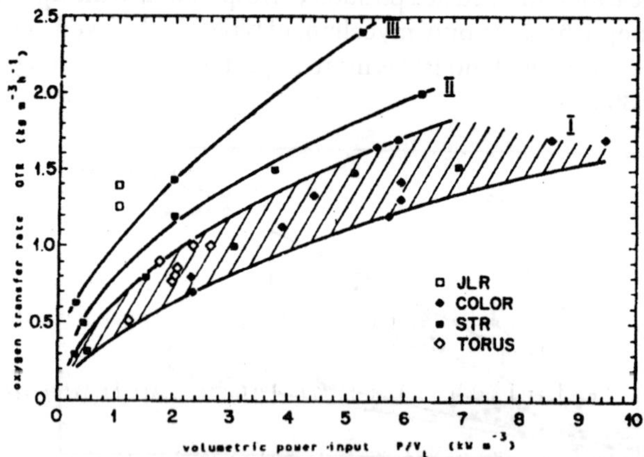

Fig. 18.16: Region I represents the oxygen transfer capacity of both propeller loop reactors and the stirred tank reactor at a gas flow rate of 0.5 vvm. Higher gas flow rates only in the stirred tank rector yields higher oxygen transfer rates: 1.0 vvm (II) and 1.5 vvm (III). The jet loop reactor (JLR) again yields good oxygen transfer at low power input.

With mycelial growth, the problem of wall growth is well known. In stirred tank reactors for instance this becomes increasingly important with decreasing reactor volume. Even at a scale of 0.3 m^3 wall growth occurs to a significant level and prohibits reliable determination of kinetic data. Here the COLOUR shows its advantages due to low mixing times and controlled flow with high turbulence. Wall growth does not occur at the scale of 0.39 m^3 and can be excluded also at 0.05 m^3 following experience with other mycelial systems.

CONCLUSIONS

The results have shown the applicability of the biological test systems and allow the evaluation of different designs. Process development is now possible with more reliability. The following three steps are recommended:

1. determination of the biological and chemical parameters in chemostat cultures
2. determination of the physical parameters with biological test systems at different process scales
3. determination of the process-specific parameters with the production system at pilot and production scale.

REFERENCES

Adler, I. and Fiechter, A. 1986. *Bioprocess Eng.*, 1, 51-59.

Fiechter, A. 1984. *Biochem. Eng/Biotechnol.*, 30, 7-60.

Hug, H., Blanch, H.W. and Fiechter, A. 1974. *Biotech. Bioeng.*, 16, 965-85.

Jaramillo, A., Adler, I. and Fiechter, A. 1986. *Chem. Ing. Tech.*, 58 (11) 914-15.

Krebser, U., Adler, I. and Fiechter, A. 1987. *Chem. Ing. Tech.*, 59 (6) 514-15.

Krebser, U., Adler, I. and Fiechter, A. 1988. *Dechema Biotechnology Conferences.* Vol. 1, pp. 123-29.

Rutte, U., Krebser, U., Adler, I. and Fiechter, A. 1987. *Proc. 4th Europ. Congr. Biotechnology.* Vol. 1., pp. 361-63

Wagner, B., Adler, I. and Fiechter, A. 1987. *Proc. 4th Europ. Congr. Biotechnology.* Vol. 1, pp. 123-26.

Yonsel, S. and Deckwer, W. D. *1990. BioEng.*, 6 (3) 12-23.

19

BIOPROCESS DEVELOPMENT

P. ETTLER AND Z. REHACEK
Institute of Microbiology, Czech Academy of Sciences, Prague, Czech Republic

Abstract. Compared to chemical reactors, there are additional requirements to be considered for the selection of suitable reactor during bioprocess development. Cultivated micro-organisms and most of their products are to a greater or lesser extent sensitive to temperature, pH and shear stress. Determination of optimal time-temperature sterilization regime also plays an important role. Many design procedures of chemical reactors can be applied in bioreactors. Some requirements, however, are specific, e.g., prevention of excessive foaming, shear sensitivity, shift from Newtonian to non-Newtonian flow properties.

Bioprocess development covers microbial strain improvement, modification of seed and production media, and alteration and changes in physical fermentation parameters for satisfaction of specific requirements of the producing strain.

Bioprocess development should be combined with the adequate strategy for successful scaling-up including the determination of oxygen uptake rate profiles for various mixing systems and bioreactor geometries. For this procedure chemical engineering and physiological criteria like P/V, K_La value, T_{mix} DOC, concentration of biochemical markers and specific enzymes activities are used.

In the beginning there was a wild strain isolated from the soil. All the hope of the biotechnologist was invested in it. To derive better production he improved it knowingly or unknowingly by repeated selection and/or mutagenic treatment, with a traditional approach or with the impact of gene technology. Industrial microbiologists have relied to a great extent on mutagenesis and screening for strain improvement, employing classical genetic approaches, exploiting sexual recombination, parasexuality, cell fusion and polyploidization (Elander, 1982; Rowlands, 1984; Bigelis, 1989).

The reasons for increased production of various substances are different. Most commonly the involved metabolic or regulation changes cause increase in the respective enzyme activity and/or inhibition of product degradation, reduction of side reactions, stimulation of substrate uptake, or improved excretion of a desired intracellular product (Kikuchi, 1980; Nakayama, 1986; Shio, 1986).

The biopress development involves–besides the improvement of the strain, seed and production media–modification of physical fermentation parameters such as temperature, pH, aeration, mixing, optimization of media sterilization, antifoaming, oxygen transfer and length of fermentation (Omstead *et al.*, 1989). The interplay between micro-organism and engineering parameters may be characterized by several features (Kuenzi, 1977):

1. cell proliferation and product formation often occur in separate phases
2. the producing micro-organisms often undergo morphological changes; high-producing mutants differ from their parent strains
3. within the strain development there is a metabolic rate difference leading to a delayed response to environmental changes
4. a network of inter-relations exists between the control variables and the behaviour of the culture described by the state variables.

In ensuring a good product recovery and purification, media composition, ionic strength, and speed of operation are the over-riding factors. During the bioprocess development, some other important elements should also be taken into account, e.g., selection of micro-organism without undesirable pigments and metabolites, modification of fermentation conditions, timing of harvesting, pH control of harvesting, temperature, flocculation, broth treatment (Aiba *et al.*, 1973; Atkinson and Sainter, 1982; Dunnill and Lilly, 1972; Darbyshire, 1981).

Undesirable interactions can be overcome by selection of the appropriate cultivation methods, batch, fed-batch, extended fed-batch or continuous operation. The selection of an alternative substrate or the use of fed-batch systems are strategies for bypassing substrate repression.

STRAIN IMPROVEMENT, INOCULUM DEVELOPMENT

During the screening of micro-organisms for industrial application the following criteria should be considered (Bull et al., 1979): nutritional characteristics, optimum temperature of cultivation, stability of genetic expression, amenability to genetic manipulation, high substrate conversion, ease of product recovery, reaction of micro-organism with equipment, etc.

The inoculum development should minimize the period of adaptation of the culture in the production medium. Inoculum must be built up in a number of stages to produce a sufficient biomass for inoculation. The length of the lag phase is affected by the size of the inoculum and by its physiological state.

The preparation of fungal inoculum is more involved than that for bacterial and yeast processes (Hockenhull, 1980), being capable of asexual sporulation of solid media and in submerged cultures (Vezina and Singh, 1975). Some fungi do not produce asexual spores and therefore an inoculum of vegetative myselium must be used (Elander, 1982; Rowlands, 1984; Bigelis, 1989; Kikuchi, 1980; Nakayama, 1986; Shio, 1986).

MEDIA DEVELOPMENT

In most cases, information on essential nutrients are available from preliminary experiments or genetic investigation (Suomalainen and Oura, 1971; Rehacek and Sajdl, 1990). In designing composition of media appropriate selection of components is necessary based both on aspects of regulatory physiological effects and on economy. The requirements of the occurring growth and production kinetics must be met. To formulate a reasonable mixture of substrates and ingredients it is necessary to understand the physiological demands of the producing strain. These could be different in the phase of the culture growth and in the phase of the products formation. The strain improvement is conditioned by a suitable modification of seed and production media, involves testing of various carbon sources, and the optimal selection of a cheap and complex nitrogen source. Biological as well as chemical and practical aspects in media preparation, standard deliveries of homogeneous raw materials are more desirable than an optimization of arbitrarily selected compounds. To identify critical parameters, intra- and extracellular parameters are used for the control of relationships in the product formation process.

Experiments investigating the possibility of application of individual ingredients are laborious, expensive and time-consuming. Media are often developed by trial and error methods. Alternatively, it is possible to use various statistical methods that allow testing of multiple variables concurrently (Box *et al.*, 1978). That optimization procedure uses the response surface methodology allowing an estimation of up to five variable interactions. Along obtained curves the optima of medium composition occur.

Nutrient medium should meet the following criteria (Stanbury and Whitaker, 1984):

- it should enable formation of maximum yields of product and/or biomass
- it should permit the maximum rate of the product biosynthesis
- there should be the minimum yield of undesired products
- it should be cheap, of consistent quality, and readily available.

Other criteria are used to select media suitable for sporulation, inoculation and substance production:

The first step of the medium formulation is based on stechiometry equations (Atkinson and Mavituna, 1983) and on knowledge of the elemental composition of a respective micro-organism. These data serve as a guide to absolute minimum quantities of C, N, P, Mg, K to be included in the initial medium. Other factors include quality of the process water, pH stability, sterilization, vitamins, biofactors and precursors.

The design of a medium is also related to the oxygen demand. The individual medium components influence the viscosity of the final broth and its subsequent behaviour during aeration and agitation. Many of the antifoams act as surface agents and reduce the oxygen transfer rate (Charles, 1978).

The calculations are focused on the major items such as macroelements that are needed in large amounts as nutrients. Trace elements are added in the inorganic or organic form, depending on the metabolic characteristics of the cell. Regulatory effects are important especially with respect to nitrogen. Organic N- compounds may be sensitive to heating and inactivation during medium sterilization.

Stability of pH can be accomplished by addition of calcium carbonate or by selection of suitable complex of organic substrates. Typical demand of nutrients for microbial growth was published by Fiechter (1973).

Information of variations of cell and product yields indicate an essential suitability of media composition or need of their significant alterations. Preliminary calculations of substrate consumption, and metabolic and yields coefficients represent the first step for future scale-up steps and simultaneously indicate the fermentation economy (Nyiri and Charles, 1977; Ratledge, 1977).

To achieve process improvement and better process control, important metabolic parameters were determined (Ryu and Hospodka, 1980; Ackerson *et al.*, 1989), e.g., specific uptake rates of C, N, O, P, S and precursor that can support the maximum production rate.

PROCESS DESIGN

During the cultivation in a nutritionally adequate medium and suitable physico-chemical environment, micro-organisms grow in a constant and regular manner. The balanced growth of the culture can be prolonged by additions of fresh nutrients and by withdrawals (Lim, 1977). In such an auto-processing system, toxic metabolites are constantly removed, an equilibrium environment is set up, and the cell division is continuous. By manipulation of the steady state growth during the bioprocess development, it is possible to exploit the physiological potential of the micro-organism, e.g., by changes in the growth environment (Pirt, 1983) provoking varying phenotypic expression of the microbial culture. During selection of the optimal bioreactor configuration, the user has to meet very many recommendations. Their major attention is paid to hydrodynamics, power requirements, and gas transfer.

BIOREACTOR DESIGN AND OPERATION

The main objective of the bioreactor design is to develop a capability to meet the demands of the interactive system in hierarchical scheme economically. During the bioreactor operation the emphasis shifts toward most efficient exploitation of the cellular machinery for optimal productivity of the process. The objectives and design variables of a general biotechnological process can be grouped into three categories:

- those related to the genetic optimization
- those related to the configurational optimization and reactor design
- those related to the process optimization and reactor operation (Oldshue and Bajpai, 1991).

A tremendous diversity exists in the types of bioreactors. One classification is based upon physical and mechanical attributes. The category of agitated bioreactors includes the forced circulation to enhance mass and heat transfer, and mixing or homogenization of suspensions. The stirred tank reactor has remained the most common type of bioreactor in use, if the large capacity of unagitated vessels used in the beverage industry is not considered. Lately, pneumatic systems have gained attention (Reuss and Bajpai, 1990).

Considerations in reactor design cover the following key factors: the requirement of sterility, rheology of broth, mass transfer, mixing, heat transfer, homogenization of suspensions and shear sensitivity of micro organism. Numerous other factors are suggested: by the interactions in the first level of biotechnological subsystem as catabolic and anabolic reactions, control of enzymic activity, degradation reactions, reaction stechiometry; in the second level transport of metabolites across membranes, development of compartments, differentiation, changes in genetic expression; in the third level population growth and production; in the fourth level bioreactor operations and upstream processing (Prokop and Votruba, 1976). The relative importance of the reactor design depends on the nature of the biological system being considered: shear effects may not be important for several microbial systems, but become crucial in the case of fragile mammalian cells. Gas-liquid mass transfer is undoubtedly a critical factor in many aerobic fermentations (antibiotics, ergot alkaloids) and in high-density recombinant bacterial or yeast systems. On the other hand it may not be important for shear sensitive cells with low growth rates (Oldshue and Bajpai, 1991; Oldshue, 1983; Oldshue, 1985; Taguchi, 1971).

A mass transfer between liquid and single cell is often not crucial and situations may change rapidly with the formation of flocs or pellets (Palotas *et al.*, 1988).

Reactor design for animal, plant or insect cells with very slow growth (compared to that of microbial cells), high shear sensitivity and damaging effect of interfaces on cellular activity must address the very stringent requirements of sterility for extended periods of operation, high shear sensitivity and oxygen supply. Reactor configurations range from classical stirred tank reactor and its modifications, air-life fermenters and hollow fibres to fluidized bed reactors. One of the main objectives of these reactor designs has been to supply oxygen minimizing or eliminating the exposure of cells to the air-liquid interface (Oldshue and Bajpai, 1991; Oldshue, 1989).

Mixing Equipment and Operation

The design of mixing equipment is complicated by diversity of systems and mixing operation. In several biosystems, the broth may undergo substantial changes in its rheology (Charles, 1978; Nienow, 1990). The mixer design is a compromise requiring proper match of different requirements.

Study of the actual process in a plant depends on finding the control mechanisms and indicating how the mixers are to be scaled up or scaled down.

Many different types of impellers can be split into two main categories: (a) with axial flow, (b) with radial flow.

A general outline of a practical pilot planting procedure can be summarized as follows: (Metz *et al.*, 1981; Van Suidam and Metz, 1981; Buckland, 1984; Oosterhuis and Kossen, 1984):

- to choose an impeller type able to satisfy the combination of flow and fluid shear required in the process
- to measure as many different process data for at least three speeds of operation
- to change D/T of the single impeller for study of effects of overall flow and fluid shear pattern
- to model the ratio of the depth of fluid to the tank-diameter in the full-scale unit of the pilot plant.

The concept of macro- and microscale shear and the procedures for identification of control factors should allow easier identification of the scale-up criteria.

In geometrically similar reactors, the scale up criteria are the following:

- circulation and blend time
- impeller tip speed
- impeller Reynolds number
- ratio of energy dissipation
- gas-liquid mass transfer coefficient for a constant volume of gas per volume of liquid per minute.

Stirred bioreactor parameters that are not biological system-specific:

- bulk fluid mixing
- air dispersion capability
- heat transfer
- fluid dynamics local to the agitator
- power input
- oxygen mass transfer coefficient.

Biological mixing performance parameters:

1. System specific parameters:
   - growth of micro-organism and secretion of product
   - damage of micro-organisms
   - morphology changes
   - O_2 and CO_2 transfer rates.

2. Parameters which may be dependent on stirring conditions:
   - quality of homogenization
   - oxygen toxicity
   - oxygen starvation
   - CO_2 concentrations
   - pH value, nutrient consumption
   - shear damage.

To compare the efficacy of performance of agitation systems it is necessary to measure the power requirements. The power represents a significant cost in running large-scale mycelial fermentations (Charles, 1978; Ettler and Marouskova, 1984). To facilitate design of agitators, system-specific biological parameters should be related to generic non-specific physical aspects of agitation–especially for viscous mycelial fermentations where complicated feedback interdependence among morphology, rheology, mixing, and transfer processes may occur.

Mass Transfer

Microbial growth is a heterogeneous catalytic reaction. The catalysts and substrates are located in different phases. Hence, substrates must pass at least one barrier in order to reach the reactive cell surface. The surface reaction occurs at a rate dependent on the substrate concentration and conditions at the surface environment. The difference between surface conditions and those in the bulk substrate phase depends on the resistance offered by the diffusion barrier, and is strongly influenced by mixing conditions and rheological properties (Charles, 1978). The mass transfer path for a gaseous component is considered in three phase diffusion model.

The conditions of motion of a gas bubble in a liquid are determined by three factors: (a) the value of the Reynolds number, (b) the shape of the bubble, (c) the interface characteristics. The shape of a gas bubble is determined by hydrodynamic forces and surface phenomena.

High pressure of gases effects microbial processes (Enfors, 1975). It increases the gas transfer rate, stability of proteins, cell divisions, energy metabolism, oxidation of SH-groups in enzymes and co-enzymes, etc. (Ollodart and Blair, 1965; Haugaard, 1955; Luck *et al.*, 1954).

Production of free radicals by oxidative enzyme systems has been suggested to be a mechanism behind oxygen toxicity under high pressure (Trager and Onken, 1988).

The reactions of oxygen and carbon dioxide occupy a central position in microbial physiology (Wimpenny, 1969; Akashi *et al.*, 1979). The optimization of culture conditions in relation to O_2 supply and CO_2 removal has played a key role in the commercial production of amino acids. Oxygen deficiency in L-glutamic acid and L-phenylalanine biosynthesis causes a decrease of the product yields (Smith and Ch. Ho, 1985). Oxygen shortage of L-aspartic acid fermentation may serve as an example of biosynthesis which occurs in oxygen-limited culture by way of PEP and oxalacetate including CO_2 fixation. CO_2 is necessary for spore germination of *Bacillus circulans, B. stearothermophilus,* for growth of *Penicillium roqueforti,* but it inhibits some enzymes of purine and L-methionine biosynthesis. At higher pressures, the appearance of swollen and stunted hyphae of *P. chrysogenum* predominated, and significant quantities of spherical or yeast-like cells were observed (Smith and Ch. Ho, 1984). Overdosing of CO_2 leads to a narcotic effect and irreversible damage of micro-organism. Rols and Goma (1989) presented new concepts to improve O_2 supply in aerobic fermentation using O_2-vectors i.e., liquids insoluble in the fermentotion media. Utilization of the O2-vectors in an emulsified form can significantly increase the oxygen transfer coefficient between gas and aqueous phase. These stated compounds are haemoglobin (Adlercreutz and Mattiasson, 1982), hydrocarbons (Mimura *et al.*, 1969), and perfluorocarbons respectively (Mattiasson and Adlercreutz, 1987).

Resistance of Micro-Organisms to Hydrodynamic Stress

Mechanical forces during stirring, pumping and aeration influence the growth of microbial culture and production of its metabolites. The sensitivity depends among others on the cell wall stability, cell morphology, physico-chemical influences, surface tension, and broth viscosity. Physiological stress in the absence of essential nutrients increases the effect of mechanical stress (Markl *et al.*, 1991).

An increase in the speed of rotation of the stirrer increases the yields of citric acid due to the improved supply of oxygen (Ujcova *et al.*, 1980). After reaching an optimum any further increase in the stirring speed rapidly reduces the citric acid production. Conditions of mixing also influence production of other organic acids (Ettler *et al.*, 1991). Similar effect was observed in the case of penicillin (Konig *et al.*, 1981). Sittig and Heine (1977) compare a switch from a soft air-lift to a jet causing the increased consumption of substrate, probably due to a higher hydrodynamic load but without a corresponding increase in the productivity of biomass.

The higher shear stress is sometimes harmful (Flaschel *et al.*, 1986), and sometimes advantageous to the process. There is difficulty in determining the direct effect of changed conditions of flow, in view of the simultaneous change in the state of aeration, etc.

Ettler *et al.* (1991) measured the flow properties *in situ* by means of sterilizable piezoelectric sensor. The effect of hydrodynamic stress gives rise to a simultaneous escape of intracellular material (Tanaka *et al.*, 1975a), microscope detection or rheometry in laminar region (Tanaka *et al.*, 1975b; Midler and Finn 1966), measuring the stress in a special jet device with nozzle under nitrogen (Markl *et al*, 1991).

The critical stirrer speed of rotation below which no damage was in evidence, decreased with increasing viscosity. The addition of CMC produced cell aggregates which did not break up even at high stirring speed (Bronnenmeier and Markl, 1982).

In the filamentous form of fungi, the rheological behaviour is highly viscous so that agitation and aeration become difficult. This is the reason why the pellet form is used Blanch and Bhaveraju, 1976).

The results stress the importance of coupling characteristics of mixing and microbial kinetics.

Measurement of circulation time distributions in different reactors and the correlations of its parameters with reactor configuration, scale and other physical and operating parameters of broth are required in order to see its impact upon the performance of the microbial system concerned. It is hoped that such an analysis would lead not only to a better understanding of operation at different scales and in different reactors, but also to a more rational approach to their design. Mixing is expected to play an important role in systems with continuous supply of sparingly soluble oxygen and slow-controlled feeding of inhibitory substrates.

Foaming

Foaming is often taken as an auxiliary aspect of the overall fermentation process. The effect of surfactants and chemical antifoam agents on the gas hold-up, the liquid phase mass transfer coefficient and the volumetric liquid phase mass transfer coefficient, are often crucial during bioprocess development, and similarly the main parameters in the design of fermenters are gas liquid mass transfer and mixing in the liquid and gas phases (Koide *et al.*, 1985; Koide *et al.*, 1976; Pandit, 1990). The value of the coefficient depends on the bubble size and the level of turbulence in the liquid phase. The specific interfacial area depends on the Sauter mean bubble diameter and the fractional gas hold-up. The bubble size in fermenters also depends on liquid phase physical properties such as liquid viscosity, liquid density, surface tension and electrolyte concentration. Large bubbles disengage easily from the liquid phase but provide little gas-liquid interfacial area. Excessively small bubbles are difficult to separate from the liquid. The presence of foamy structures in fermentations can offer some advantages (Kumar, 1987), but on the other hand excessive foaming can cause problems with non homogeneity, sterility, and nutrient limitation owing to the low liquid hold-up. To control fermentation foams, efficiency in suppressing the foam is not the only consideration (Evans and Hall, 1971). Quite often, foaming tendency changes during the course of fermentation, resulting from changed microbial activity and depending on original medium composition and ingredients. The function of substrate concentration, pH value, sterilization time, ionic strength, bulk viscosity, etc., together with quality of introduced inoculcum are described in Szarka and Magyar (1969). The desirable features of antifoam agents are (Solomons, 1967): fast knock-down, long-lasting action, high efficiency, non-toxic to micro-organisms and macro-organisms, no complications in down-stream processing, low flammability, cheapness, no negative effect on oxygen transfer rates.

From the technological point of view, the foam influences the filling volume of the vessel, makes it difficult to carry out the process under sterile conditions, and complicates the operation of mechanically mixed bioreactors. Foam formation is in some cases connected with the flotation of micro-organisms resulting in lowering of the active biomass content and decrease of process productivity.

REFERENCES

Abbott, B.J. and Clamen, A. 1973. *Biotechnol. Bioeng.*, 15, 117.

Ackerson, M.D., Clausen, E.C. and Gaddy, J.L. 1989. *Appl. Biochem. Biotechnol.*, 20, 511.

Adlercreutz, P. and Mattiasson, B. 1982. *Eur. J. Appl. Micro-biol. Biotechnol.*, 16, 165.

Aiba, S., Humphrey, A.E., Millis, N.F., 1973. Recovery of fermentation products. *Biochem. Eng.* 2nd edition, Academic Press, New York. pp. 346-92.

Akashi, K., Shibai, H. and Hirose, Y. 1979. *J. Ferm. Technol.*, 57, 321.

Atkinson, B. and Sainter, P. 1982. *J Chem. Tech. Biotechnol.*, 32, 100

Atkinson, B. and Mavituna, F. 1983. *Biochem. Eng. and Biotechnol. Handbook.* Macmillan, London. pp. 114-203.

Bigelis, R., 1989. In H.J. Rehn and G. Reed, (eds), *Biotechnology.* VCH Verlagsgesellschaft, Weinheim. Vol. 7b, pp. 230-69

Blanch, H.W. and Bhavaraju, S.M. 1976. *Biotechnol. Bioeng.*, 18, 745.

Box, G.E., Hunter, W.G. and Hunter, J.S. 1978. *Statistics for Experimenters.* J. Wiley and Sons Inc. New York, pp. 510-39.

Bronnenmeir, R. and Markl, H. 1982. *Biotechnol. Bioeng.*, 24, 553.

Buckland, B.C. 1984. Biotechnology 2, 875.

Bull, A.T., Ellwood, D.C. and Ratledge, C. 1979. *Soc. Gen. Microbiol. Symp.*, 29, 668.

Charles, M. 1978. In T.K. Ghose, A. Fiechter and N. Blakebrough (eds), *Adv. Biochem. Eng.* Springer Verlag, Berlin. Vol. 8, pp. 1-62.

Darbyshire, J. 1981. In A. Wiseman and E. Horwood (eds) *Large Scale Enzyme Extraction and Recovery. Topics in Enzyme and Fermentation Biotechnology.* Chichester. Vol. 5, pp. 147-86.

Dunnill, P. and Lilly, M.D. 1972. *Biotechnol. Bioeng. Symp.*, 3, 97.

Elander, R.P. 1982. In V. Krumphanzl, B. Sikyta and Z. Vanek (eds), *Overproduction of Microbial Metabolites.* Academic Press, London. pp. 351-69.

Enfors, S.O. 1975. *Report Chem. Centre Technical Microb.*, University of Lund, Sweden.

Ettler, P., Marouskova, L. 1987. *Proc. III Europ. Congr. Botechnol.*, Munich. p. 314.

Ettler, P. 1990. *Coll. Czech. Chem. Commun.*, 55, 1730.

Ettler, P., Fort, I. and Kobliha, V. 1991. *Biotechnol. Techn.*, 11.

Ettler, P., Martinkova, L., Ujcova, E. and Seichert, L. 1991. *Folia Microbiol.*, 36, 493.

Evans, J.I. and Hall, M.J. 1971. Proc. *Biochem.*, 6, 4.

Fiechter, A. 1973. *Chem. Ing. Technol.*, 45, 2.

Flaschel, E., Raetz, E. and Renen, A. 1986. *Proc. Int. Conf. Bioreactor Fluid Dynamics,* Cambridge. p. 173.

Haugaard, N. 1955. *Proc. Underwater Physiol. Symp. Nat. Acad. Sci.* Washington, USA. pp. 8-12.

Hockenhull, D.J. 1980. In J.E. Smith., D.R. Berry and B. Christiansen, (eds), *Fungal Biotechnology.* Academic Press. pp. 1-24.

Kikuchi, M. 1980. *Biotechnol. Bioeng.*, 22 (1) 195.

Koide, K., Hayashi, T., Sumino, K. and Iwamoto, S. 1976. *Chem. Eng. Sci.*, 31, 963.

Koide, K., Yamazoe, S. and Harada, S. 1985. *J. Chem. Eng.*, 18, 287.

Konig, B., Seewald, C., Schugerl, K. 1981. *Eur. J. Appl. Microbiol. Biotechnol.*, 12, 205.

Kossen, N.W.F. and Oosterhuis, N.M.G. 1985. In H.J. Rehm and G. Reed (eds), *Modelling and Scaling up of Bioreactors.* Verlagsgesellschaft, Weinheim. Vol. 2, pp. 571-605.

Kuenzi, M.T. 1977. *Process Design and Control in Antibiotic Fermentations.* Proc. FEMS. Basel.

Kumar, R. 1987. In B.D. Kulkarini (ed), *Proc. Int. Chem. React. Eng. Conf. II.* Wiley Eastern Ltd., New Delhi. pp. 190-207.

Lim, H.C. 1977. *Biotechnol. Bioeng.*, 19 (3) 425.

Luck, H. 1954. *Schweiz. Z. Allg. Pathol. Bakteriol.*, 17, 106.

Markl, H., Bronnenmeier, R. and Wittek, B. 1991. *Int. Chem. Eng.*, 31 (2) 185.

Mattiasson, B. and Adlercreutz, P. 1987. *Trends Biotechnol.*, 5, 250.

Metz, B., De Bruijn, E.W. and van Suidam, J.C. 1981. *Biotechnol. Bioeng.*, 23, 149.

Midler, M. and Finn, R.K. 1966. *Biotechnol. Bioeng.*, 8, 71.

Mimura, A., Kawano, T. and Kodaira, R. 1969. J. Ferment. Technol., 47, 229.

Mitard, A. and Riba, J.P. 1986. *J. Appl. Microbiol. Biotechnol.*, 25, 245.

Nakayama, K. 1986. *Prog. Ind. Microbiol.*, 24, 3.

Nienow, A.W. 1990. TIBTECH, 8, 224.

Nyiri, L.K. and Charles, M. 1977. *Ann. Rep. Perm. Proc.*, 1, 365.

Oldshue, J.Y. 1983. *Fluid Mixing Technology.* McGraw-Hill, New York. p. 192.

Oldshue, J.Y. 1985. *Biotechnol. Adv.*, 3, 219.

Oldshue, J.Y. 1989. *Chem. Eng. Prog.*, 5, 5.

Oldshue, J.Y. and Bajpai, R.K. 1991. In A. Prokop, R.K. Bajpai and S. Ch. Ho. (eds), *Recombinant DNA Technology and Applications.* McGraw-Hill Inc., New York. pp. 460-76.

Ollodart, R. and Blair, E. 1965. *J. Am. Med. Assoc.*, 191, 736.

Omstead, M.N., Kalan, L. and Buckland, B.C. 1989. In W.C. Campbell (ed), *Fermentation Development and Process Improvement.* Springer, New York. pp. 33-54.

Oosterhuis, M.M.G. and Kossen, N.W.F. 1984. *Biotechnol. Bioeng.*, 26, 546.

Palotas, J., Seres, P., Hogye, I. and Polya, K. 1988. *Proc. 2nd Int. Symp.* OMP. Ceske Budejovice, p. 97.

Pandit, A.B. 1990. *Adv. Biochem. Eng.*, 45.

Pirt, S.J. 1983. *J. Chem. Tech. Biotechnol.*, 33B (3) 137.

Prokop, A. and Votruba, J. 1976. *Folia Microbiol.*, 21, 58.

Ratledge, G. 1977. *Ann. Rep. Per. Proc.*, 1, 49.

Rehacek, Z. and Sajdl, P. 1990. *Ergot Alkaloids. Chemistry, Biological Effects, Biotechnology,* Elsevier, Amsterdam. pp. 215-25.

Reuss, M. and Bajpai, R.K. 1990. In C. Schugerl (ed), *Biotechnology in a Comprehensive Treatise,* Verlag Chemie, Weinheim. p. 240.

Rols, L. and Goma, G. 1989. *Biotech. Adv.*, 7, 1.

Rowlands, R.T. 1984. *Enzyme Microb. Technol.*, 6, 3.

Ryu, D.D.Y. and Hospodka, J. 1980. *Biotechnol. Bioeng.*, 22, 289.

Shio, 1. 1986. *Prog. Ind. Microbiol.*, 24, 188.

Sittig, W. and Heine, H. 1977. *Chem. Ing. Tech.*, 49, 595.

Smith, M.D. and Ch. Ho, S. 1984. *Proc. ACS Ann. Meeting,* Philadelphia.

Smith, M.D. and Ch. Ho, S. 1985. *J. Biotechnol.,* 2, 347.

Solomons, G.L. 1967. *Proc. Biochem.,* 2 (10) 47.

Stanbury, P.F. and Whitaker, A. 1984. *Principles of Fermentation Technology.* Pergamon Press, Oxford. p. 74.

Suomalainen, H. and Oura, E. 1971. In A.H. Rose and J.S. Harrison, (eds), *The Yeast.* Academic Press, London. pp. 136-45.

Szarka, L. and Magyar, K. 1969. *Biotechnol. Bioeng.,* 11, 701.

Taguchi. H. 1971. Adv. Biochem. Eng., 1, 43.

Tanaka, H., Mizuguchi, T. and Ueda, K. 1975a. *J. Ferm. Technol.,* 53, 35.

Tanaka, H., Takahashi, J. and Ueda, K. 1975b. *J. Ferm. Technol.,* 53, 18.

Trager, M. and Onken, U. 1988. *Dechema Biotechnol. Conf. 1.* VCH Verlagsgesell. p. 199.

Ujcova, E., Fencl, Z., Musilkova, M. and Seichert, L. 1980. *Biotechnol. Bioeng.,* 22, 237.

Van Suidam, J.C. and Metz, B. 1981. *Biotechnol. Bioeng.,* 23, 111.

Vezina, C. and Singh, K. 1975. In J.E. Smith and D.R. Berry (eds), *The Filamentous* Fungi. Academic Press, London. Vol. 1, pp. 158-92.

Viesture, U.E., Kristapsons, M.Z. and Levitans, E.S. 1982. In A. Fiechter (ed), *Adv. Biochem. Eng.* Springer Verlag, Berlin, Heidelberg. Vol. 21, pp. 169-224.

Wimpenny, J.W.T. 1969. *Proc. 19th Symp. Soc. Gen. Microbiol.* Cambridge Univ. Press. p. 164.

Worgan, J.T. 1986. *Prog. Ind. Microb.,* 8, 73.

20

PROCESS CHARACTERISTICS OF MECHANICALLY AGITATED BIOREACTORS

I. FORT, P. DITL AND V. VANECEK
Czech Technical University, Prague, Czech Republic

INTRODUCTION

For project, design, and production of the mechanically agitated bioreactor (see Fig. 20.1), it is necessary to calculate the following characteristics:

- biological characteristics–type of micro-organisms and morphology, culture medium, respiration rate, temperature of microbial suspension, etc.
- process characteristics (see complete reference list)–critical shaft frequency of revolution for flooding/ loading transition, power input of gassed and ungassed impeller (different types of high-speed impellers, see Figs. 2-4), heating and/or cooling time of the culture medium in the course of sterilization and cooling or heating of microbial suspension, etc.
- design characteristics (Ruzicka, 1969)–shaft diameter, thickness of walls and baffles or other internals, design of impellers, thickness of jacket or coil, etc.

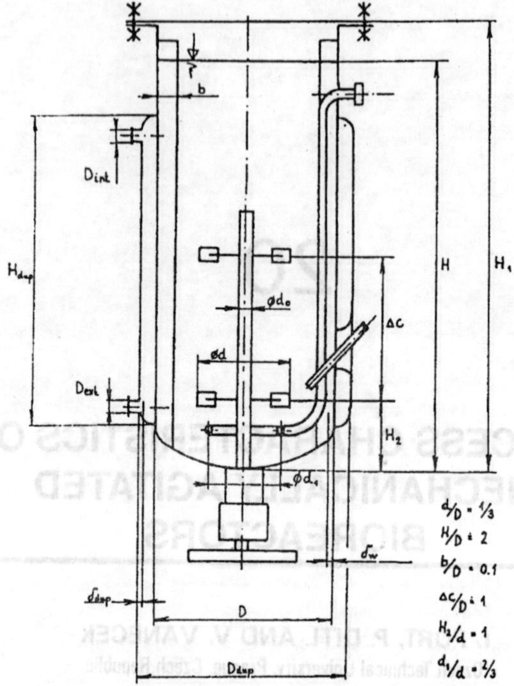

Fig. 20.1: Sketch of mechanically agitated bioreactor with two standard turbine impellers on the same shaft.

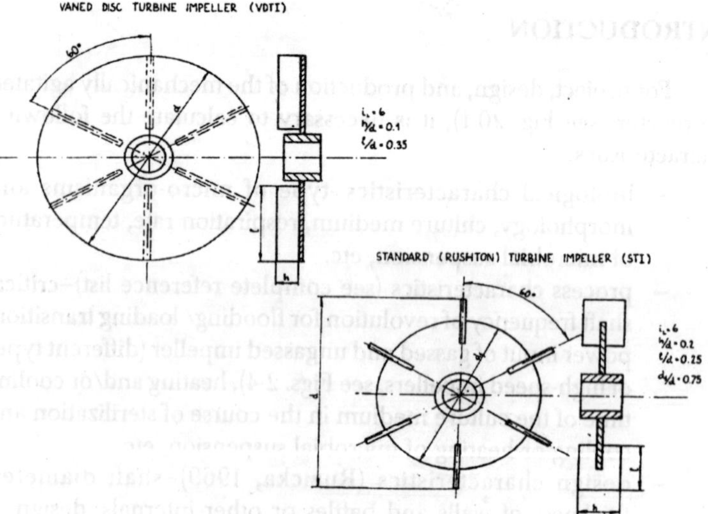

Fig. 20.2: Radial high-speed impellers

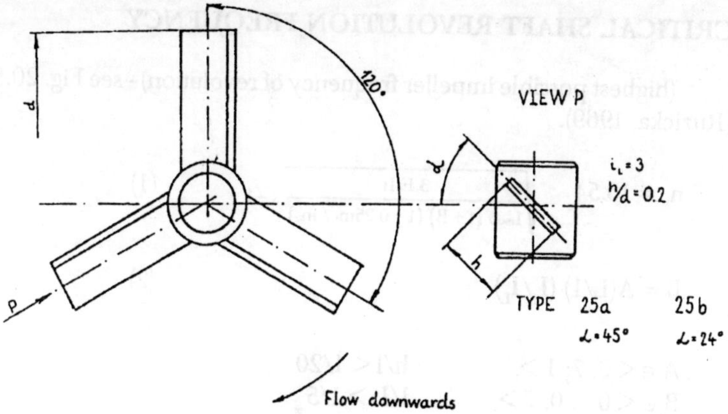

IMPELLER WITH SIX INCLINED PLANE BLADES (ISIPB)

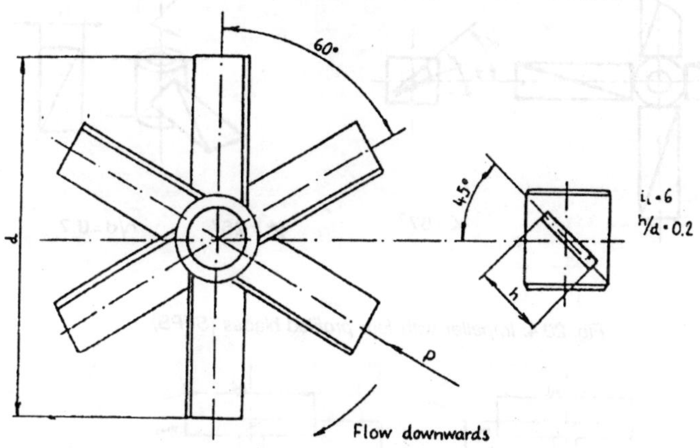

Fig. 20.3: Axial high-speed impellers

The biological characteristics of bioreactors serve as the basic level for the operations considered and are usually decisive when final results of bioreactors are required. Process characteristics of cultivation decide the yield of the fermentation process and are usually an important background for the scaling-up of the bioreactors.

Design characteristics include results of stress analysis necessary for manufacturing the bioreactor body, shaft, impeller, and internals.

This contribution contains selected process characteristics of bioreacctors equipped with different types of high-speed mixers: high-shear (radial) impellers and low-shear (axial) impellers.

CRITICAL SHAFT REVOLUTION FREQUENCY

(highest possible impeller frequency of revolution) - see Fig. 20.5 (Ruzicka, 1969).

$$n_k = 16.54 \sqrt{\frac{3\, E\, I_1}{l_{mo}3\,(1+B)\,(1+0.25 m_s / m_o)}}\ s^{-1} \qquad (1)$$

$$B = A(L/l)\,(I_l/I_L) \qquad (2)$$

$A \in <0.7;\ 1>,\qquad h/l < 1/20$

$B \in <0\ ;\ 0.7>,\qquad l/L > 1/5$

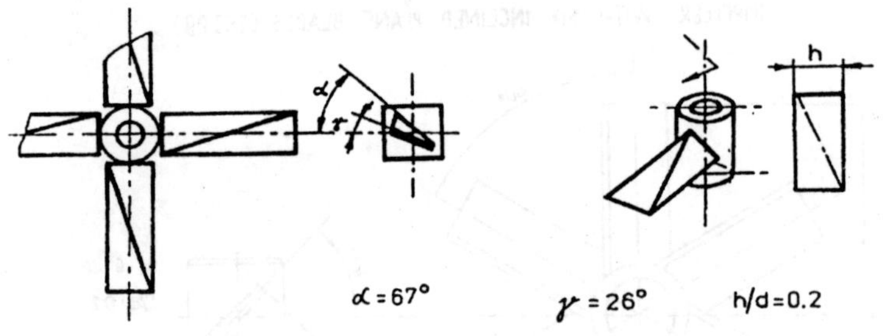

Fig. 20.4: Impeller with four profiled blades (SFPB)

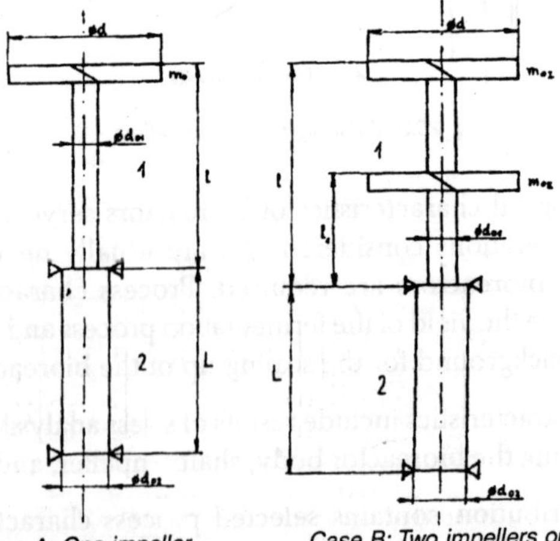

Case A: One impeller *Case B: Two impellers on the same shaft*

Fig. 20.5: Sketch of shaft with bearings and one or two impellers.

l, L length of correspond. Parts of the shaft (m)
I_l moment of shaft cross-section intertia in part l (m^4)
I_L moment of shaft cross-section inertia in part L (m^4)
E moduli of elasticity of the shaft (Pa)
m_o mass of the impeller (kg)
m_s mass of the shaft (kg)

FREQUENCY OF REVOLUTION FOR FLOODING/ LOADING TRANSITION

Machon *et al.* (1985) (lowest possible frequency of revolution)- see Fig. 20.6.

$$Fr_m^{\ A} = k\,Kp \qquad (3)$$

$$Fr_m = n^2 d/g \quad \text{Froude number for mixing} \qquad (4a)$$

$$Kp = \dot{V}g/n_{fl}\,d^3 \quad \text{Flow rate number} \qquad (4b)$$

n_{fl} impeller frequency of revolution at flooding/loading transition (s^{-1})
$\dot{V}g$ gas volumetric flow rate ($m^3\ s^{-1}$)
d impeller diameter

CONSTANTS OF EQN (3) FOR DIFFERENT IMPELLER GEOMETRIES

Impeller (See Figs. 20.2-20.4) Lower	Upper	A	k	Note	
ISIPB	-	1.645	6.217	Pumping	downwards
STI	-	1.0	2.70		
VDTI	-	1.0	2.668		
STI	STI	1.21	2.231		
STI	ISIPB	1.53	3.173	Upper impeller is pumping	upwards
STI	ISIPB	1.12	6.61	Upper impeller is pumping	downwards

H/D = 1 or 2, $d/D = H_2/D = 1/3$, $\Delta\ c/D = 1$
four radial baffles, $b/D = 0.1$, $Re_m > 5.0\times10^3$

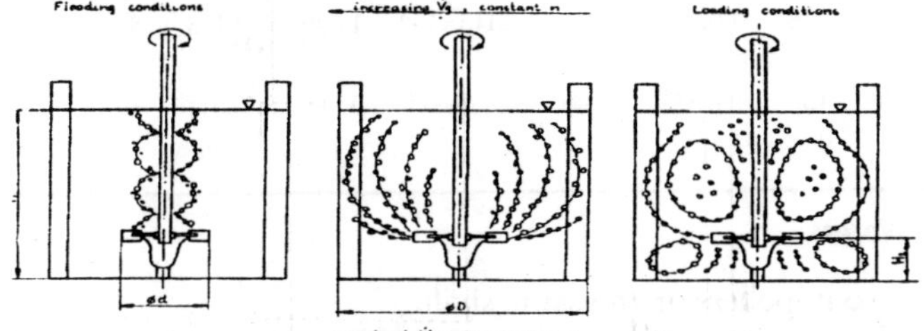

Fig. 20.6: Definitions of flooding and loading conditions in gas -liquid mixing (radial high-speed impeller).

POWER INPUT OF UNGASSED IMPELLER

Vavro *et al.* (1988)

$$Po = \left[\left(\frac{P1}{Re_m}\right)^{p2} + (Re_m)\, p_4\, ^{p_3} (Re_m)\, p_5 + p_6\right]^{1/p_2} \quad (5)$$

Where

$Re_m \quad \in < Re_{m\,1}; Re_{m2} >$

$Po \quad = P/(\rho n^3 d^5)$ Power number (6a)

$Re_m \quad = n\, d_2\, \rho/\mu$ Reynolds number for mixing (6b)

ρ density of agitated batch (kg/m³)

μ (apparent) viscosity of agitated batch (Pa s)

Notes:

Laminar flow regime at mixing

$Po = p_1/Re_m \qquad Re_m < Re_{ml}$ (7a)

$Po = (P_6)^{1/p2} \qquad Re_m > Re_{m2}$ (7b)

CONSTANTS OF EQN (5) FOR DIFFERENT IMPELLER GEOMETRIES

Impeller	p_1	p_2	p_3	p_4	p_5	p_6	Re_{ml}	Re_{m2}
ISIPB	64	0.836	-5.50	0.617	0	1.56	10	1×10^4
STI	73.1	1.595	-6.51	3.24×10^{-5}	1	13.44	5	1×10^4
DTI	0	1	0	0	0	2.6		5×10^4
ITIPB ($\alpha = 45°$)	47.5	1.074	-0.214	3.145×10^{-5}	1.609	1.108	8	2×10^4
ITIPB ($\alpha = 24°$)	40	0.640	-0.906	0.365	0	0.511	10	1×10^4
SFPB	0	1	0	0	0	0.96		5×10^4

Notes:

Two impellers on the same shaft:

$P_{total} = P_{upper} + P_{lower}, \qquad \Delta c/d \geq 2.0$ (8)

Power Input of Gassed Impeller (See Fig. 20.7) (Oldshue, 1989; Machon *et al.*, 1985)

$$\frac{1}{(1-P_g/P)} = K + q \frac{(P_g/P)^{0.5}}{Kp} \qquad (9)$$

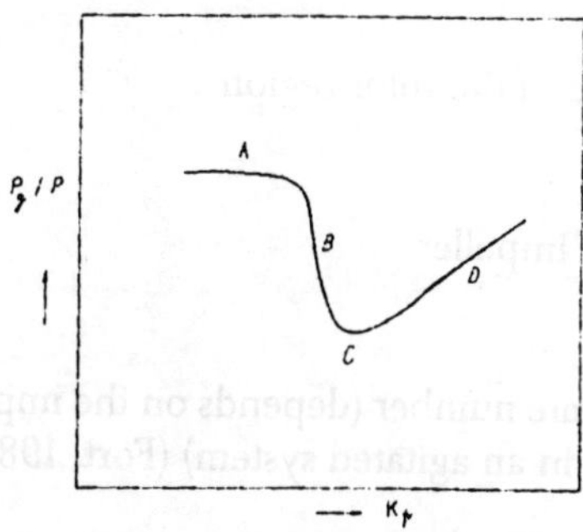

Fig. 20.7: Dependence of gassed/ungassed impeller power input ratio and flow rate number kp. A indirect loading, B direct loading, C flooding/loading transition, D flooding conditions.

CONSTANTS OF EQN (3) FOR DIFFERENT IMPELLER GEOMETRIES

Impeller Lower	Upper	k	q	Note
STI	-	1.43	0.029	
VDTI	-	1.41	0.037	
STI	STI	1.44	0.066	
STI	ISIPB	1.46	0.042	Upper impeller is pumping upwards
STI	ISIPB	1.62	0.055*	Upper impeller is pumping downwards
STI	ITIPB ($\alpha = 45°$)	1.43*	0.055*	Upper impeller is pumping upwards
STI	ITIPB ($\alpha = 24°$)	1.43*	0.042*	
STI	SFPB	1.43*	0.055*	

* Estimation from semi-empirical theory.

$H/D = 1$ or 2, $d/D = H_2/D = 1/3$, $\Delta c/D = 1$
four radial baffles, $b/D = 0.1$, $Re_m > 5.0 \times 10^3$

Equation for calculation of the flooding/loading transition in agitated system with an axial high-speed impeller (Machon and Fort, 1992)

bubble buoyancy drag force of the bubble in the stream streaking from the impeller rotor region

$(1/6)\ g\ \pi\ d_b\ (\rho_l - \rho_g)$

$\stackrel{!}{=} \varphi\ \rho_l\ (1/4)\ \pi\ d_b^{\ 2}\ (1/2) w^2$ (10)

db = diameter of the bubble

Average velocity of liquid streaking from the impeller rotor region

$w = \dot{V}p/S$ (11)

Circular cross-section of the rotor region

$S = \pi\ d^2/4$ (12)

Pumping capacity of impeller

$\dot{V}p = Cp\ n\ d^{-3}$ (13)

Cp = impeller flow rate number (depends on the impeller geometry, its size, and position in an agitated system) (Fort, 1986)

Then,

$w = (4\ Cp/\pi)\ (nd)$ (14)

and bubble Reynolds number

$Re_b = w\ d_b/\upsilon = 4\ Cp\ n\ d_b\ d/\pi\ \upsilon_l$ (15)

υ_l = kinematic viscosity of liquid

Experimentally determined constants

Cp = 1.0 six blade impeller with inclined plane blades (x= 45°)ISIPB (see Fig. 20.3) $d/D = H_2/D = (1/3)$, $Re_m > 1.0 \times 10^3$

$\varphi = 0.44\ {}_re_b > 1.0 \times 10^3$

$\dot{V}_b - 1.378\ V_g^{\ 1.2}/g^{0.6}$, single spherical bubbles (Fort, 1986)

or

$d_b = 0.874\ \dot{V}_g^{\ 0.4}$ (16)

Constants of the pilot plant agitated system

$\rho_l = 1000\ kg/m^3$, water at 20°C

$\rho_g = 1.205\ kg/m^3$, air at 20°C

$\upsilon_l = 1.308 \times 10^{-6}\, m^2/s$, water at 20°C

$d = 0.096$ m (impeller diameter)

Critical impeller frequency of revolution at flooding/loading transition (for the system considered)–see Figu. 20.7.

$$n_{fl} = (1.99 \times 10^3\, d_b)^{0.5} \qquad (s^{-1};\, m) \qquad (17a)$$

$$n_{fl} = 41 \times 68\, \dot{V}_g^{\;0.2} \qquad (s^{-1};\, m^3 s^{-1}) \qquad (17b)$$

Flooding/loading transition in a system with an axial high-speed impeller: conclusions

Critical frequency of impeller revolution at flooding/loading transition can be calculated when equilibrium of forces effecting on the gas bubble in the region below the rotating impeller is considered.

Experimentally obtained and theoretically calculated conditions of the flooding/loading transition were found to be in fairly good agreement.

Axial high-speed impellers are suitable for gas dispersion, though they are more sensitive with respect to the volumetric gas flow rate than radial high-speed impellers (e.g., Rushton turbine).

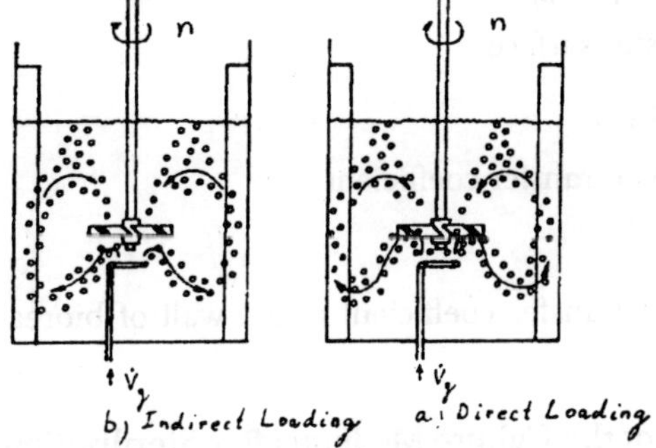

Fig. 10.8: Definitions of flooding and loading condition in gas-liquid mixing (axial high-speed impeller).

Scaling-up Criteria for Mechanically Agitated Bioreactors

A Constant gas volumetric flow rate per batch volume

$$V_{g,\, \underline{\underline{M}}}/V_M = \dot{V}_{g,\, I}/V_I \qquad (18)$$

B1 Constant specific impeller power input

$$P_{g', \underline{\underline{M}}}/V_{\underline{\underline{M}}} = P_{g', I}/V_I \quad (19a)$$

B2 Constant impeller tip speed

$$\pi\, d_M\, n_{\underline{\underline{M}}} = \pi\, d_{\underline{\underline{I}}}\, n_{\underline{\underline{I}}} \quad (19a)$$

(Complex (mycelial) microbial suspensions)

Index 'M' . . . pilot plant equipment

Index 'I' . . . industrial equipment.

Heating Time of the Culture Medium before Sterilization (Saturated Water Vapour)

(McCabe and Smith, 1976)

$$J_h = \frac{Q}{k\, A\, \Delta t_{LMTD}} \quad (20)$$

Heat necessary for heating of the culture medium

$$Q_h = m_{st}\, r = \rho\, V\, c_p\, (t_s - t_o) \quad (21)$$

Logarithmic mean temperature difference

$$t_{LMTD} = \frac{(t_V - t_o)\,(t_V - t_s)}{\ln \dfrac{(t_V - t_o)}{(t_V - t_s)}} \quad (22)$$

Heat transfer surface

$$A = \pi D\, H_{DUP} \quad (23)$$

Overall heat transfer coefficient

$$k \doteq \alpha_M \quad (24)$$

α_M = heat transfer coefficient at the wall of bioreactor in an agitated batch

Cooling Time of the Culture Medium after Sterilization (Cooling Water)

$$\frac{(t_s - t_c)}{(tf - tc)} = \frac{\dot{m}_c\, c_{p,\, c}\,[1 - \exp(-N_A\, \tau_c)]}{\rho\, V\, cp} \quad (25)$$

Number of thermal units

$$N_A = A\, k / (_{mc}\, c_{p',\, c}) \quad (26)$$

Heat time needed for cooling of culture medium

$$Q_h = m_{c\,cp}, \dot{c}\,(t_s - t_f)\, \tau_{\underline{c}} = \rho\, V\, c_p\, (t_s - t_f) \qquad (27)$$

$$k = [(1/\alpha_M) + (\delta_{DPU}/\lambda_w) + (1/\alpha_c)]^{-1}\ (W\ m^{-2}) \qquad (28)$$

Heat transfer coefficient in an agitated batch

$$Nu_M = (\alpha_M\, D/\lambda) = 0.76\, Re_M^{\ 2/3}\, Pr^{\ 1/3}\, (\mu/\mu_w)\ 0.14 \qquad (29)$$

$Pr = c_p/\lambda$ = Prandtl number

Notes:

Equation (29) is valid for one or two high-speed impellers on the same shaft if Δ c/D>1.

Heat transfer coefficient in cooling water $\alpha_{\underline{c}}$ depends on the design of the cooling jacket of the bioreactor.

CONCLUSIONS

- Critical shaft and/or impeller frequency of revolution when vibrating and when impeller is flooded serve as an estimation of the working interval of frequencies used.
- Power input of ungassed impeller serves for selection of the driving motor of the bioreactor.
- Power input of ungassed impeller serves for scaling up of the fermentation process from the pilot plant to the industrial production unit.
- Heating and cooling times of the culture medium before and after sterilization serve for determination both of the heating and cooling media and of the lengths of both periods.
- All the above-mentioned process characteristics of mechanically agitated bioreactors are important background for the project of the control of the fermentation process.

EXAMPLE

It is necessary to design a mechanically agitated fermenter for submersion production of citric acid by means of the mould *Aspergillus niger.* Following scale-up, criteria were found for the industrial aerated fermenter:

- constant impeller speed π dn = 3.5 m/s
- constant superficial velocity of gas in the fermenter

$w_o = 4\, \dot{V}_g / \pi D^2 = 1.5$ mm/s

Main geometric characteristics of the industrial fermenter:

volume V= 30 m^3, standard Rushton disc turbine d/D = 1/3, cylindrical vessel with flat bottom and four baffles b/D = 1/10.

There are two possibilities for the design of the industrial fermenter, either with the ratio H/D = 1.0 (one standard Rushton turbine impeller), or with its ratio H/D = 2.0 (lower impeller standard Rushton turbine impeller, upper impeller–six blade impeller with inclined-plane blades–$\alpha = 45^o$, h= 0.2 d).

Decide which arrangement is better from the point of view of its energy requirement and calculate the total power output of the whole apparatus (the mechanical efficiency of the impeller gear box η_M = 0.90). Take into consideration both the work of compression under isothermal conditions with the efficiency of compressor η_c = 0.40, the gassed turbine impeller power input and ungassed inclined-plane blades impeller input. Calculate for both cases whether impeller is flooded or not.

Physico-chemical properties of liquid phase:

density ρ_l = 1050 kg/m^3, viscosity μ_l = 50 mPa s.

Total atmospheric pressure, air, Pa = 1.0 × 10^5

Pa, and its density ρg = 1.100 kg/m^3.

Frequency of impeller revolution on flooding is given by (Mikulkova *et al.*, 1967):

$$(Fr_m = Kp_g)\ (d/D)^{3.43} = 6.11 \times 10^{-2},\ Re_m > 1.0 \times 10_4 \quad \text{(i)}$$

Where the Froude number for mixing is defined as

$$Fr_m = n^2\, d/g \quad \text{(ii)}$$

and the flow rate number for gas is defined as

$$Kp_g = V_g/nd^3 \quad \text{(iii)}$$

Gassed impeller power input P_g can be calculat[illegible] follows (Strek,1977):

$$1/\left[1-(P_g/P)\right] = K + q\sqrt{(P_g/P)/Kpg},\ Re_m > 1.0 \times 10_4 \quad \text{(iv)}$$

where parameters in eqn (IV) above are given in the following table for the standard turbine impeller (d/D = 1/3):

K= 1.44, q = 0.0660 (two standard turbine impellers on same shaft)

K+ 1.55, q = 0.0381 (one impeller on the shaft).

K= 1.46, q = 0.0620 (two impellers on same shaft).

Ungassed power input P can be calculated from the power number

$$P_o = \frac{P}{P_1 n_3 d^5} \qquad Re_m > 1.0 \times 10^4 \qquad (v)$$

where for the given impellers its values are as follows (Czechoslovak Standard, 1969):

standard turbine impeller $P_o = 5.6$

impeller with six inclined-plane blades $P_o = 1.7$ (d/D = 1/3)

REFERENCES FOR THE EXAMPLE

Czechoslovak Standard 6910. 1969. VUCHZ - CHEPOS, Brno.

Mikulcova, E., Kudrna, V. and Vlcek, J. 1967. *Sci. Papers of Prague Inst. Chem. Technolog.*, K1, 167.

Strek, F. 1977. *Mixing and Mixing Equipment*. SNTL. Prague. In Czech.

Solution of the Problem

Impeller diameter		d
Gas flow rate		$\dot{V}_g$
Ungassed impeller power input		P
Gassed impeller power input		P_g
Theoretical compressor power input		P_t
Gassed impeller power output	P_M]	Total power
Compressor power output	P_C]	output P_t

Numerical Solution

Subscript i = A - ratio H/D = 1

Subscript i = B - ratio H/D = 2

A geometric variables and simplexes

i= A - H/D = 1

$V= (\pi/4)\, D^2 H \Rightarrow V= (\pi/4) D^3$ (A)

$D= [(4/\pi)\, V]^{1/3} = [(4/\pi)\, 30]^{1/3} = 3.36$ m

$d/D = 1/3, \Rightarrow d = 1.12$ m (B)

$b/D= [1/10 \Rightarrow b = 0.34$ m

$i = B - H/D = 2 \Rightarrow V = (\pi/4)\, D^2\, 2D = (\pi/2)\, D^3$ (C)

$D = [(2/\pi)\, V]^{1/3} = [(2/\pi) 30]^{1/3} = 2.67$ m

$d/D = 1/3 \Rightarrow d = 0.89$ m (D)

$b/D = 1/10 \Rightarrow b = 0.27$ m

B gas flow rate

$w_o = 4\, \dot{V}_g / (\pi D_2) \Rightarrow \dot{V}_g = (\pi/4)\, w_o\, D^2$ (E)

$i = A \text{-} \dot{V}_g = (\pi/4)\, 0.0015\; 3.36^2 =$

$= 0.0133\ m^3\ s^{-1} \doteq 0.8 m^3\ min^{-1}$

$i = B \text{-} \dot{V}_g = (\pi/4)\, 0.0015\; 2.67^2 =$

$= 0.0084\ m^3\ s^{-1} \doteq 0.5\ m^3\ min^{-1}$

C impeller frequency of revolution

$n = \pi d\, n/\pi d$

$i = A - n = 3.5/ (\pi\, 1.12) = 1.00\ s^{-1} = 60\ min^{-1}$

$i = B - n = 3.5/ (\pi\, 0.89) = 1.25\ s^{-1} = 75\ min^{-1}$

D frequency of impeller revolution on flooding

$(Fr_m/Kp_g)\, (d/D)^{3.43} = 6.11 \times 10^{-2}$ (G)

$(n_{fl}^2\, d)/g\, (n_{fl}\, d^3)/\dot{V}_g\, (d/D)^{3.43} = 6.11 \times 10^{-2}$ (Ga)

$n_{fl} = [6.\ 11 \times 10^{-2}\, (\dot{V}_g/d^4)]^{1/3}\, (D/d)^{1.143} = i = A\text{-}$ (H)

$n_{fl} = [6.11 \times 10^{-2}\, (0.0133\ 9.81/1.\ 12^4)]\ 1/331.143$

$= 0.60\ s^{-1} = 36\ min^{-1} < n$

$nfl = [\ 6.11 \times 10^{-2}\, (0.084\ 9.81/0.89^4)]\ 1/3_3 1.143$

$= 0.70\ s^{-1} = 42\ min^{-1} < n$

E ungassed power input

$P = Po\ \rho_I\ n^3\ d^5.$ (I)

$Po = Po_I + Po_{II}\ (H/D = 2, \Delta\ c/d \geq 2)$ (J)

$i = A\text{-}\ Po_I = 5.6$

$P\ \ = 5.6\ 1.050\ 1.0^3\ 1.12^5 = 10.362\ W$

$i = B\ \text{-}\ Po_I = 5.6, Po_{II} = 1.7$

$Po\ \ = 5.6 + 1.7 = 7.3$

$(\Delta\ c/d \geq 2 \Rightarrow \Delta c \geq 1.88\ m)$

$P = 7.3\ 1.050\ 1.25^3\ 0.89^5 = 8.350\ W$

F impeller Reynolds number

$Re_m = n\ d^2\ \rho_l/\mu_l$

$i = A\ \text{-}\ Re_m = 1.00\ 1.12^2\ 1.050/0.050$

$= 2.63 \times 10^4$

$i = B\ \text{-}\ Re_m = 1.25\ 0.89^2\ 1.050/0.050$

$= 2.07 \times 10^4$

G gassed impeler power input

$[1\text{-}(P_l/P)] = K + q\ \sqrt{(P_g/P)}\ /\ Kp_g$

$1/[1\text{-}(P_g/P)] = K + (q/Kp_g)\ \sqrt{P_g/P}$

$i = A\ \text{-}\ Kp_g\ \dot{V}_g/(n\ d^3) = 0.0133/(1.0\ 12^3) =$

$= 9.467 \times 10^{-3}$

$1/[1\text{-}(P_g/P)] = 1.55 + \sqrt{P_g/P\ (0./0381/0.009467)}$

$P_g/P\ \ = 0.806 \Rightarrow P_g = 10.362\ 0.806 =$

$= 8.352\ W$

$i = B\ \text{-}\ Kp_g = \dot{V}_g/(n\ d^3) = 0.0084/\ (1.25\ 0.89^3) =$

$= 0.532 \times 10^{-3}$

$1/\ [1\text{-}(P_g/P)] = 1.46 + \ \sqrt{P_g/P}\ \ (0.0620.0/0.009532)$

$P_g/P\ \ = 0.867 \Rightarrow P_g = 8.350.67 =$

$7.240\ W$

H theoretical compressor power input

$P_t = P_g 1 \dot{V}_g 1 \, Wt$ (N)

$W_t = p1 \dot{V}_g 1 \ln (p_2/p_1)$ (O)

$Vg_1 = 1/P_{g1}$ (P)

Subscript 1 ... input (atmospheric) conditions

Subscript 2 ... output conditions of the compressor

$P_2 = p_1 + \rho_1 gH$ (Q)

$i = A - v_{g1} = 1/1.1 = 0.909 \; m^3/kg$

$W_t = 1.0 \; 10^5 \; 0.909 \times$

$\times \ln [(1.0 \; 10^5 + 1.050 \; 9.81 \; 3.36) / (1.0 \; 10^5)]$

$W_t = 2.70 \; 10^4 \; J \; kg^{-1}$

$P_t = 1.1 \; 0.0133 \; 2.70 \; 10^4 = 395 \; W$

$i = B - v_{g1} = 0.909 \; m^3/kg$

$W_t = 1.0 \; 10^5 \; 0.909 \times$

$\times \ln [(1.0 \; 10^5 + 1.050 \; 9.81 \; 3.36)/(1.0 \; 10^5)]$

$W^t = 3.98 \; 10^4 \; J/kg$

$P_t = 1.1 \; 0.0084 \; 3.98 \; 10^4 = 368 \; W$

I impeller power output

$P_M = P_g / \eta_M$

$i = A - P_M = 8.352/0.90 = 9280 \; W$ (R)

$i = B - P_M = 7.240/0.90 = 8044 \; W$

J compressor output

$P_c = P_t / \eta_c$ (S)

$i = A : P_c = 395/0.4 = 988 \; W$

$i = B : Pc = 368/0.4 = 920 \; W$

K total power output

$P_t = P_M / P_c$ (T)

$i = A - P_t = 9.280 + 988 = 10.268 \; W = 10.3 \; kW$

$$i = B - P_t = 8.044 + 920 = 8.964\ W = 9\ kW$$

Measurement of Kinetic Energy of Turbulence by Means of Piezoelectric Transducer

The output signal of the piezotransducer, see Figure 20.9, placed in a multiphase medium, is a result summation of interactions between a sphere of the transducer and each phase of the medium. Output signal is directly proportional to the sum of the specific energy of the medium including all the phases present:

$$e = K \sum_{j=1}^{N} \rho_j \varphi_j v_j{}^2 \tag{1}$$

where

N number of phases in the medium

j index of the j th phase

ρ_j density of j th phase

φ_j volumetric ratio of j th phase

v_j local velocity of j th phase.

For turbulent quasi-stationary flow, we can consider that:

$$\varphi_j = \bar{\varphi} j + \varphi j' \tag{2a}$$

$$= \bar{v}j + vj'$$

and

$$\bar{\varphi}_j' = \bar{v}j' = 0, \qquad =0, \quad j = 1, 2, \ldots N \tag{3a}$$

$$\varphi_j 'v_j' = \varphi_j 'v_j'^2 \quad = 0, \qquad j = 1, 2, \ldots N \tag{3b}$$

$$\rho_j \quad = \rho_{j'}, \qquad j = 1, 2, \ldots N \tag{3c}$$

We can then rearrange eqn (1) to apply eqns (2a) and (2 b) and obtain:

$$e = K \sum_{j=1}^{N} \rho_j (\varphi_j + \varphi_j')(v_j +^2 v_j') \tag{4}$$

and finally

$$e = K \sum_{j=1}^{N} \rho_j (\varphi_j v_j^2 + \varphi_j^2 v_j^2 + \varphi_j^2 v_j^2 + \varphi_j^2 v_j^2 + \varphi_j^2 v_j v_j' + \varphi_j^2 v_j^2) \tag{5}$$

Let us introduce the mean value of the specific kinetic energy of the turbulent flow e as the time-averaged value

$$\bar{e} = (1/T) \int_0^T e\,(t)\,dt \qquad (6)$$

After time integration of the right-hand side of eqn (6) with application of eqns (3), we have

$$\bar{e} = \sum_{j=1}^{N} \rho_j \overline{(\varphi_j V^2 + \varphi_j V^2)} \qquad (7)$$

or after rearranging

$$\bar{e} = \sum_{j=1}^{N} \rho \overline{(\varphi_j v^2)} + \sum_{j=1}^{N} \rho_j \overline{(\varphi_j v^2)} \qquad (8)$$

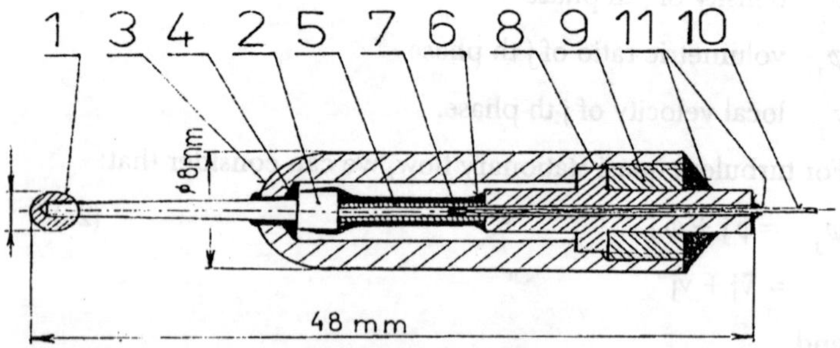

Fig. 20.9: Piezotransducer

The first term on the right-hand side of eqn (8) chracterizes the contribution of the mean velocity and the second term the contribution of the fluctuation to the mean specific energy of the turbulent flow of the multiphase medium. Thus the second term characterizes the mean specific energy of turbulence

$$\overline{e'} = \sum_{j=1}^{N} \rho_j \overline{\varphi_j v_j^2} \qquad (9)$$

and it can be determined from the known fluctuating signal measured by the transducer.

For single-phase flow (j=1), we can arrange eqn (9) into the form

$$\overline{e'} = K \rho \overline{v'^2} \quad (10)$$

and for two-phase gas-liquid flow (j=1, g), we can write

$$\overline{e'} = K \left(\rho_g \overline{\varphi_g} \, \overline{v_g'^2} + \rho_l \overline{\varphi_l} \, \overline{v_l'^2} \right) \quad (11)$$

Take into consideration that

$$\rho_l / \rho_g \approx 10^3 \quad (12a)$$

and

$$\overline{\varphi_g} < 0.1 \quad (12b)$$

we can neglect the first term on the right-hand side of eqn (11), and we have finaly

$$\overline{e'} = K \rho_l \overline{\varphi_l} \, \overline{v_l'^2} \quad (13)$$

Notes

Proportionality constant K can be calculated from calibration of the transducer by means of the turbulence velocity field (e.g., LDA).

With respect to the validity of eqns (12), it is possible to consider that eqn (3b) is valid at least for the gas-liquid system.

Integration period T in eqn (6) depends on the character of turbulence in an investigated system (e.g., for mechanically agitated system it is about 2-3 minutes).

REFERENCES

Clift, R., Grace, J.R. and Weber, M.E. 1978. *Bubbles, Drops and Particles.* Academic Press, New York.

Fort, I. 1986. In V. W. Uhl and J.B. Gray (eds), *Mixing. Theory and Practice.* Academic Press, New York. Vol. III.

Machon, V. *et al.*1985. Proc. *5th Europ. Conf. Mixing,* Wurzburg. BHRA Cranfield. pp. 105-65.

Machon, V. and Fort, I. Coll. Czech. Chem. Commun. In press.

McCabe, W.L. and Smith, J.C. 1976. *Unit Operations of Chemical Engineering.* McGraw-Hill, New York.

Oldshue, J. 1989. Chem. Eng. Prog., May, 33-42.

Ruzicka, J. 1969. *Machinery*, 19, 494-96. In Czech.

Vavro, K. *et al.* 1988. *Proc. Czech. Conf. Mixing*, Znojmo. Czechoslovak Society of Engineers, Brno. pp. 150-55.

Zlokarnik, M. *Chem. Eng. Prog.*, 86 (42), 62.

21

SOFTWARE FOR BIOPROCESS CONTROL, AUTOMATION AND ARTIFICIAL INTELLIGENCE

B. SONNLEITNER
Institute for Biotechnology, ETH Zurich, Switzerland

INTRODUCTION

Improvement of biotechnology hardware (bioreactors, peripheral equipment, sensoric and analytic devices) is one aspect of the promotion of reliability of bioprocesses. The other, equally important factor is biotechnology software. This comprises methodologies of, and algorithms for, process operation and control, signal validation, data evaluation and reduction, observation and recognition of the physiological state, its regulation and dynamics. The methodologies must be suitable for automation; this requirement is directly obvious for all control applications but at the same time favours the improvement of the scientific quality of academic research.

SOFTWARE CONCEPT CONCEPT

The software concept developed and routinely operated at ETH relies on high modularity and is hierarchically well structured (-even reflected by the hardware istallations; Fig. 21.1).

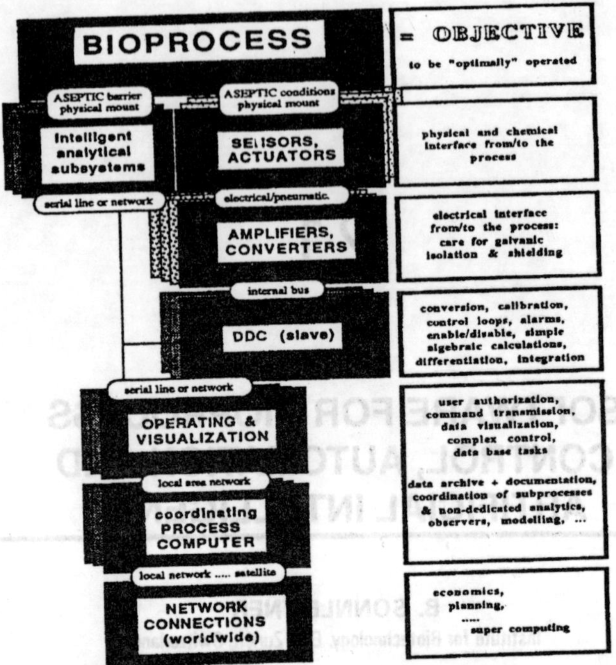

Fig. 21.1: Overview of the hierarchy of bioprocess automation components: structure and function aspects. Left-hand side shows the single modules (hardware components) as well as their connections; right-hand side shows the major tasks and key elements. The two lower elements need not necessarily be present for a sound automatic bioprocess operation. The principally high flexibility in extending and (inter) connecting various components is indicated by shadowed boxes.

The front end component, a direct digital control unit, functions as a low level service for:

- galvanic isolation, conversion and calibration of all signals
- operating simple and mostly single closed control loops (e.g., T, pH, P_{O2}..)
- safety supervision (e.g., master control, enable/disable switches)
- simple step sequence control (e.g., sterilization, refill procedures, etc.).

This front end should be buffered with an uninterruptable power supply, provide at least 12 bit conversion resolution, and must have a mathematical co-processor. It should contain a great variety of basic logical and mathematical functions as firmware and communicate with other systems via efficient and rapid serial lines (at least RS232, better Ethernet or similar).

The DDC's cycle time goes as low as 100 ms. Although bioprocesses have a characteristic duration of hours or days, the low cycle time is decisive for successful controllers of physical and chemical state variables (especially pressure, gas and liquid fluxes, etc.).

This requirement is often ignored by control engineers who have little experience with bioprocesses. Data are transferred to and from the next level, the supervisory process computer, every 15s. This value was found to be a compromise and sufficient for every day work, but it must be reduced significantly during controller tuning or in rapid transient biological experiments.

The responsibilities of the supervisory level are:

- data representation (it is, therefore, a graphic workstation)
- highly organized data storage after selective reduction; relational database)
- on-line calculation of models (for data evaluation or prediction, e.g., kalman filter)
- on-line recognition of typical patterns and states
- decision-making (in order to start sequences as feed, harvest, pulse or shift)
- on-line calculation of more difficult control laws
- communication with the front end as well as dedicated intelligent analytical subsystems.

Fig. 21.2 gives an overview of programmes and data structures used and exploited on this computer level. It is important to realize that the data/information flux is a closed loop.

The programmes serving intelligent analytical subsystems (such as FIA, G, HPLC, MS, ...) are scheduled by the same central installer task and resemble the shown concept to high degree. The data link and synchronization between these layers occurs on the level of shared memory areas and database.

Depending on the load of active programmes and on the frequency of their execution this may result in a CPU load in the range from 15 to 100% of a powerful workstation (like a VAXStation 3100). This feature calls for an urgent need to implement analytical subsystems as 'intelligent' ones. This means that a considerable portion of data evaluation and validation must be performed by the subsystem autonomously; see Fig. 21.3.

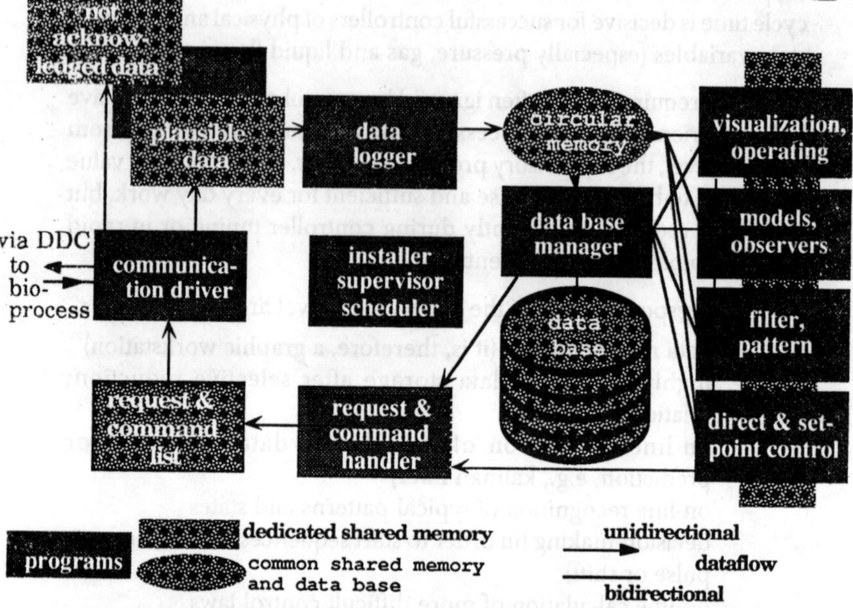

Fig. 21.2: Basic concept of distributed programmes and shared memory areas on a supervisory station. Only data flow is indicated but not control interactions for inter-task communication and synchronization. This type of universal implementation with currently 10 to 12 programmes and 6 to 8 common shared memory areas is running on 3 independent workstations at ETH.

A communication driver programme handles all requests and commands to, and responses and other messages from, the front end controller; since data transfer is always embedded in a distinct protocol, the driver has to translate transmitted data into 'plain text' data and vice versa. Information to the slave component is taken from a sorted list of requests and commands, information from the slave is tested for plausibility and written to either the 'plausible data' section or into another section to be further treated by special recovery programmes. This driver runs with high priority and adds a time stamp to every data item. A data logger programme transfers plausible data into a sorted circular memory section that holds the newest history of plausible data and places a pointer to the most recent data record. It is the only programme with write access to this memory section and has also high priority to make new data rapidly available to other programmes. Any user-specific programme (some examples are indicated) may retrieve recent data directly from the circular buffer (with unlimited read only access rights). A special programme performs data reduction and writes selected data into the permanent database; this database manager is the only programme with write access to the database. Although other programmes may retrieve data from the database directly, it is convenient to let the database manager also handle read requests for those programmes. These programmes may need to communicate with each other; they might even run on external CPUs. To close the loop of data flow, a request and command handler sorts the information from controller programmes, eventually adds information from the database, and places the requests according to their priority in the outgoing list. There is a central supervisor programme that organizes and schedules all the single programmes but does not directly participate in the data handling.

Biospecific software has to consider as many data as can be obtained about the process. The independent variable in most biological experiments is time.

An important exception are the steady-state experiments (dilution rate). It is nowadays established that most experimental studies of physiology are based on the investigation of the trajectory of a dependent variable with respect to the independent variable (time course, x-D diagram) because, obviously, much more information can be derived from these data than from single analytical values (e.g., yield or conversion at the end of a culture).

An important determinant in physiological studies is the necessary frequency of analyses (if they are not continuous signals), of data storage and evaluation. Considerable information about *dynamics* may be lost or overlooked when this aspect is underestimated.

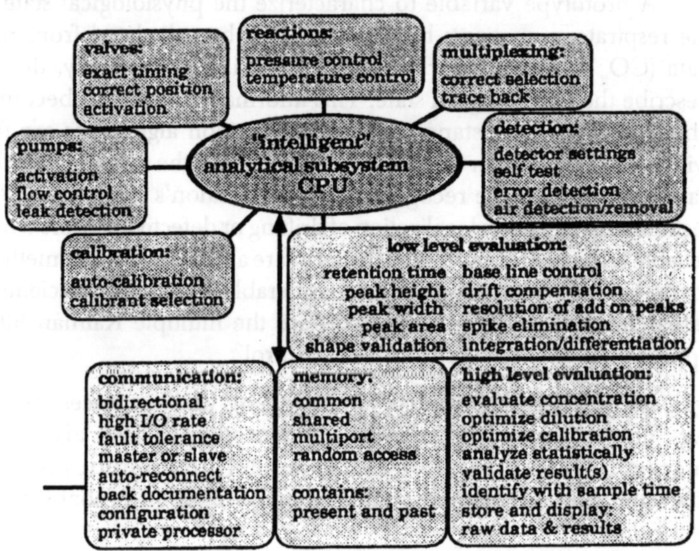

Fig. 21.3: Principal overview of tasks to be performed by an intelligent analytical subsystem. In the top part, there are the low level control actions necessary to keep the instruments working autonomously. In the middle and bottom parts, there are the key features that make such instruments intelligent. Finally, they should be capable of transmitting validated, finalized results to a supervisory system either on their own initiative or on request.

Fig. 21.4 gives an illustrative example for batch cultures, and Fig. 21.5 for spontaneously synchronized chemostat cultures of *Saccharomyces cerevisiae.*

It is necessary to use more frequent, or better still continuous, on-line (= automated) techniques which neither disturb the system nor waste too much of the sample.

Some of the signals available from on-line sensors are difficult to calibrate or to interpret *per se*, i.e., when isolated from other information. If, however, these signals are used in combination with others, they provide some further useful information to characterize a cultivation. Exploitation of all three dimensions (1: value of signal, 2: time, 3: individual variables, i.e., signals produced by different sensors and analytical equipment) of the on-line and off-line information, i.e., working with patterns of (time) trajectories, yields more information than just the sum of single, isolated trajectories.

A prototype variable to characterize the physiological state is the respiratory quotient: but this can only be calculated from raw data (CO_2 and O_2 balances) which *per se*, i.e., individually, do *not* describe the physiological state. This information gain also becomes obvious when, for instance, pattern recognition algorithms are fed with a varying number of parallel sensor signals: the more signals the easier and safer is the recognition of a cultivation's state. The latter technique even allows localization of drifting or defective sensors when reliable reference (or prototype) patterns are available. It is one method for on-line validation of signals, of considerable importance in scientific research. Well working alternatives are the multiple Kalman filter approach, neural networks or fuzzy control.

Several different models for describing various process states, e.g., normal operation, infection, or sensor drift, have also been successfully applied. Test of the model probabilities allows discrimination on-line and identification of the most probable state of a culture.

An interesting and promising application of pattern recognition is shown in Fig. 21.6. The starting point there is the observation that the signal of P_{O2} drifts with time during a continuous culture of *S. cerevisiae* (which oscillates synchronously). It is not usually possible to decide on the basis of this individual signal alone whether the drift is due to a sensor drift or due to a drift of the population (i.e., a biological drift). One needs further information which must be provided by other sensors. If there are historical data as reference available, an experienced person (i.e., an expert) can subjectively distinguish between both possibilities. However, it might take considerable time until the problem is identified as such. In contrast, if an automated algorithm checks all the time whether there are deviations to reference data in any respect likely to occur, the problem can even be solved before being recognized in manual operation.

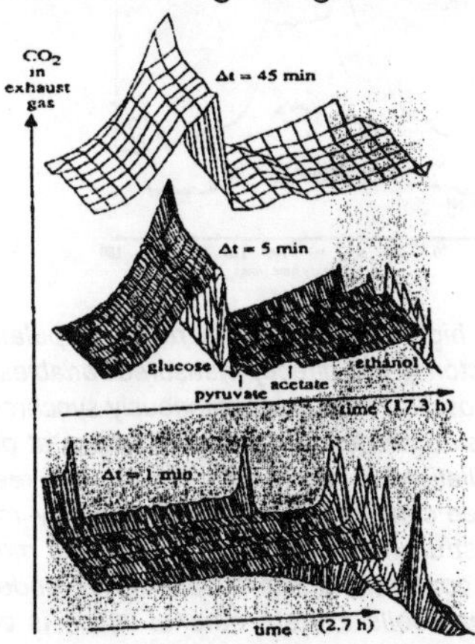

Fig. 21.4: Importance of high data density for the quantitative characterization of microbial cultures visualized for a series of batch cultures with Saccharomyces cerevisiae on chemically defined minimal medium. All 3 subplots show the CO_2 concentration in the exhaust gas, a continuous measurement, but plotted in different time resolution (Δt; as if were either an off-line or a discontinuous on-line analysis). Time increases from left to right; first batch is the series in the back, following batches are added towards the front. The first batch had been inoculated with 5% YEPD reculture, the second batch was conducted by replacing 95% of the culture with defined medium, and so on. Unknown components for the complex medium obviously do influence the growth characteristics in very low concentrations: the exponential character of growth throughout the glucose phase and the growth rate thereafter.

Top: data available every 45 min: the growth phase on glucose can easily be distinguished from the subsequent phase on ethanol. In the intermediate diauxic lag phase, little activity is obvious, if any. The exponential character of growth cannot be confirmed unequivocally.

Middle: data available every 5 min: additionally, activities in the diauxic lag phase become obvious: after depletion of glucose, 2 other substrates are metabolized with CO_2 as a by-product formed (other analyses reveal that the intermediately excreated pyruvic and acetic acids are reconsumed as indicated). Growth on glucose is not exponential over the entire range (shoulder before maximal CO_2 production). At the end of the culture, some non-reproducible activities can be deduced (peaks).

Bottom: zoomed view of the last part of the growth phase on ethanol (indicated by shaded area) with data available every 1 min: the last activity turns out to be reproducibly of very short duration (we have indications that residual acetate is consumed aft-ethanol depletion).

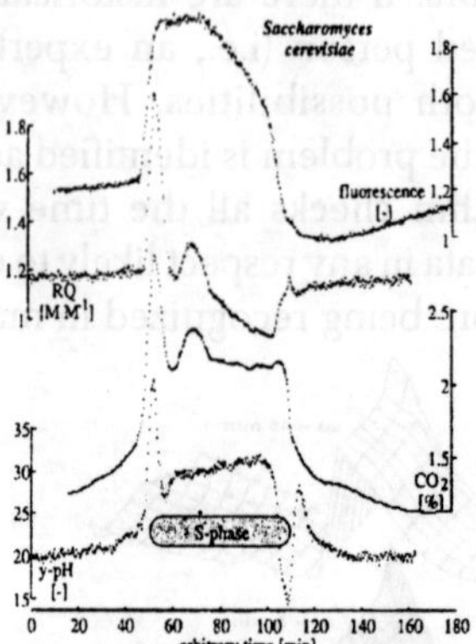

Fig. 21.5: Importance of high data density to reveal population dynamics and fast metabolic regulation: trajectories of 2 directly measured variables and 2 software sensors (calculated data) during 1 oscillation of a spontaneously synchronized chemostat culture of Saccharomyces cerevisiae on defined glucose medium at dilution rate D= 0.13 h^{-1}. The fine structure of oscillations is highly reproducible and resolved here with a time separation of 15 s (between consecutive dots). Unfortunately, mechanistic explanations for all the individual fine structures cannot be given at the moment. (Flow cytometric measurements, however, confirm that one subpopulation undergoes the cell cycle S-phase, as indicated. These cells excrete ethanol which is consumed by the other subpopulation; data not shown.)

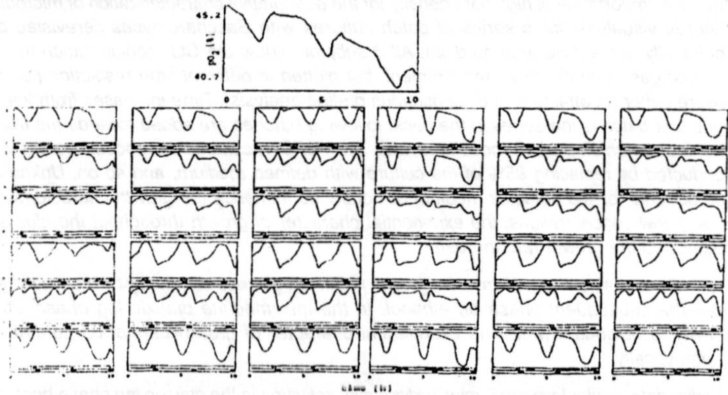

Fig. 21.6: Obviously drifting P_{O2} signal (top) and result of pattern recognition algorithm (bottom): scalar products (ordinates) are plotted vs time (abscissae) for six different patterns constructed from five on-line signals out of a set of six (in each column another one is omitted); comparison with six different reference patterns (rows) characterizing a certain phase of the synchronous oscillation of S. cerevisiae in continuous culture. In the third column, the scalar product is not subject to a temporal shift-down: the P_{O2} signal has been omitted when constructing the actual data pattern vector. Therefore, the P_{O2} signal must be identified as no longer reliable and the temporal development of the yeast population does not significantly differ from the historical reference (i.e., biology is OK but the P_{O2} sensor becomes defective). From Locher et al., 1990.

Administrative and co-ordinating tasks are established on the highest level. Among the most important duties are the service for operating system, programmes and backup, as well as documentation, data archiving, off-line calculations, organization of recipes or connection to special services (e.g., super computers, libraries, worldwide networks). If mutiplexed analyzers are used in a larger plant (servicing more than one reactor), the co-ordinating computer has the additional task of redistributing the (confidential) data to their origins (ie the respective front end or supervisory computers).

CONCLUSIONS

Automation of process control has, in our experience, improved both the reliability of single processes and the useful operating time (the yield) of the individual plants. There are strong indications that the production of artifacts has been greatly reduced. The reproducibility of biological experiments is nowadays drastically increased and no longer a dream.

The artificial intelligence of bio-process computers needs to be strongly developed. We have to date satisfying experience with the recognition of typical states or phases of a cultivation, and we are able to evaluate tupical responses of a culture to disturbances and transients by using (simple) models, but the experienced, well trained experimenter is still very important. Computers and algorithms are just excellent tools.

REFERENCES

Locher, G., Sonnleitner, B. and Fiechter, A. 1990. Pattern recognition: a useful tool in technological processes. *Bioproc. Eng.*, 5, 181-87.

Locher, G., Sonnleitner, B. and Fiechter, A. 1991. Automatic bioprocess control. 2. Implementation and practical experiences. *J. Biotechnol.*, 19, 127-43.

Locher, G., Sonnleitner, B. and Fiechter, A. 1991. Automatic bioprocess control. 3. Impacts on process perception. *J. Biotechnol.*, 19, 173-91.

Sonnleitner, B., Locher, G. and Fiechter, A. 1991. Automatic bioprocess control. 1. A general concept. *J. Biotechnol.*, 19, 1-18.

Sonnleitner. B. 1991. Quantitation of microbial metabolism. Ant v Leeuwenhock. submitted.

22

INDUSTRIAL APPLICATIONS OF AGITATED BIOREACTORS

P. SEICHTER
Research Institute of Chemical Equipment, Brno, Czech Republic

LIST OF SYMBOLS

Ax	specific energy consumption	$kWh\ kg^{-1}$
a'	specific interfacial area (gas-liquid)	m^{-1}
C_{DF}	concentration of defoaming agent	$kg\ m^{-3}$
C_L	dissolved oxygen concentration	$mol\ m^{-3}$
c^*	equilibrium oxygen concentration in liquid	$mol\ m^{-3}$
D	vessel diameter	m
D_L	diffusion coefficient	$m^2\ s^{-1}$
d	agitator diameter	m
g	acceleration due to gravity	$m\ s^{-2}$
H_o	liquid filling height	m
K_H	Henry's constant	$Pa\ m^3\ mol^{-1}$
KC	relative cL-value (kC=cL/c*)	-
K_L	mass transfer coefficient	$m\ s^{-1}$
K_{O2}	constant for oxygen ($K_{O2} = M_{O2}\ p\ p_{O2}^{-1}\ K_H^{-1}$)	-
K_{RQ}	respiration quotient	-

$\dot{M}_{O2}$	specific oxygen transfer rate	mol m^{-3} h^{-1}
m_{O2}	molecular weight of oxygen ($mO2 = 32$)	kg mol^{-1}
n	rotational speed of agitator	s^{-1}
P	power input	W
Pr_{Vo}	productivity of fermenter	kg m^{-3} h^{-1}
p	pressure	Pa
$\dot{V}_G$	gas flow rate	m^3 s^{-1}
$\dot{V}_L$	liquid flow rate	m^3 s^{-1}
o	liquid volume in fermenter	m^3
vvm	specific aeration rate ($vvm = 60\,V_G/V_o$)	min^{-1}
w_D	gas velocity in sparger orifice	m s^{-1}
w_S	superficial gas velocity	cm s^{-1}
y	molar ratio of oxygen	-
R_{em}	Reynolds number ($R_{em} = n\, d^2\, \upsilon^{-1}$)	
Sc	Schmidt number ($Sc = \upsilon\, D_L^{-1}$)	
We_M	Weber number ($We_M = n^2\, d^3\, \rho_L\, \sigma^1$	
α	exponenet in eqn (15)	-
β	exponent in eqn (16)	-
υ	kinematic viscosity	m^2 s^{-1}
ρ	density	kg m^{-3}
σ	surface tension	kg s^{-2}

Subscripts

1 inlet, 2 outlet,–mean, G gas, L liquid, O2 oxygen, D in gas sparger

For power input P: F total in fermenter, G aerated mixer, H for pumping, IS isothermal expansion

Scale-up of bioreactors has brought many problems, mostly resulting in the degradation of fermentation process parameters with large-scale apparatus. There could be several reasons for this, though the main factors resulting in these undesirable changes seem to be the differences in transport phenomena characteristics in small-scale and industrial equipment (Kossen and Oosterhuis, 1985). The change in conditions is further reflected in some alterations of the immediate micro-organism neighbourhood with subsequent decrease in their production ability.

Transport processes taking place in bioreactors are closely connected with the following practical problems which have to be solved when designing the apparatus:

- momentum transfer: mechanical energy dissipation, its intensity and distribution, circulation times in a batch, mechanical defoaming, residence time distribution (with continuous times)
- heat transfer: cultivation heat removal
- mass transfer: oxygen transfer from gas to liquid phase, carbon dioxide transfer from liquid to gas phase, blending (time) inside liquid phase, and so on.

Oxygen transfer rate (OTR) is usually the key problem for aerobic fermenter design. When referring to very slow changes of cultivation variables, the oxygen transfer can be considered as a steady-state process.

A schematic description of gas passage through the liquid phase is shown in Fig. 22.1 (left-hand side).

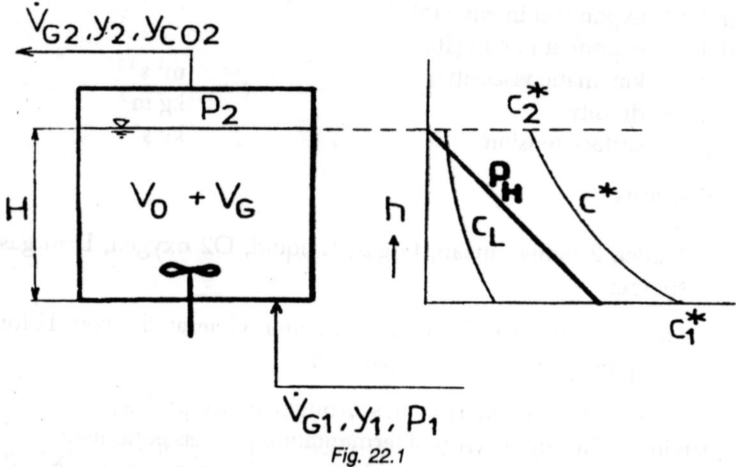

Fig. 22.1

Assuming that the inlet and outlet gas flow rates are the same (i.e., for $K_{RQ} \doteq 1$)

$$\dot{V}_{G1} = \dot{V}_{G2} \tag{1}$$

transferred oxygen amount per unit of volume and time, $\dot{M}_{O2}$, is equal to

$$\dot{M}_{O2} = 3600\,(\dot{V}_{G1}/V_o)\,(\rho_{O2}/m_{O2})\,(y_1 - y_2) \tag{2a}$$

or

$$\dot{M}_{O2} = 3600\,(\dot{V}_{G1}/V_o)\,(c_{1^*}\text{-}c_{2^*})/K_{O2} \tag{2b}$$

where y and c* are oxygen molar ratios in gas and equilibrium oxygen concentration in liquid, respectively. Simultaneously

$$\dot{M}_{O2} = 3600\,k_L a\,(\bar{c}^*\text{-}c_L) \tag{3}$$

The determination of mean value $\bar{c}^*$ depends on the way gas bubbles go through the fermentation broth. Limiting cases are represented by

- perfect mixing,
  where: $\bar{c}^* - c_L = c_2^* - c_L$ (4a)

- piston flow,
  where: $\bar{c}^* - c_L = \dfrac{c_1^* - c_2^*}{\ln \dfrac{c_1^* - c_L^*}{c_2^* - c_L^*}}$ (4b)

Since perfect mixing of gas phase in a large industrial aerobic bioreactor is hardly obtainable, piston flow of bubbles seems to be much more realistic. In the same situation, the effect of hydrostatic pressure on oxygen solubility cannot be neglected and the driving force of oxygen absorption can be determined according to both dissolved oxygen concentration (DOC) and pressure profiles (see Fig. 22.1 right hand side). To solve the problem, the following simplifying assumptions have been included:

- system is monodisperse and non-coalescent
- system is homogeneous from the point of view of interfacial area and gas hold-up
- DOC is proportional to the c*-value in a given point of the batch ($c_L/c^* = K_C$)
- mass transfer coefficient is constant.

On the basis of the above-mentioned assumptions, the relationship between volumetric mass transfer coefficient, k_La and OTR, M_{O2} has been derived (Seichter, 1987):

$$k_L a = \frac{\dot{M}_{O2} \ln\,(y_1 p_1 / y_2 p_2)}{3600\,c_1^*\,(1 - K_C)\,(1 - y_2 p_2 / y_1 p_1)}$$

The magnitude of k_La, more easily available from experiments than interfacial area itself, in agitated bioreactors is dependent, above all, on

- intensity of mechanical agitation
- gas flow rate
- properties of the liquid being mixed.

Typical correlations for the relationship are as follows:

Van't Riet (1972):

$$k_L a = f_1 [P_G/V_O, w_S ...] \quad (6)$$

Nishikawa *et al.* (1981):

$$k_L a D^2/D_L = f_2 [Re_M, Sc, We_M ...] \quad (7)$$

Henzler (1982)

$$(k_L a/w_S)(\upsilon^2/g)^{1/3} = f_3 [(P_G/V_O) P_L \upsilon_g] \quad (8)$$

Our experimental work aimed at proving the use of eqns (6-8) for steady-state fermentations was performed on two continuous processes of single cell protein (SCP) production in the VUCHZ laboratories at Brno. The $k_L a$-values were evaluated by means of eqn (5). The laboratory fermenter, 50 l in volume, is presented in Fig. 22.2. The apparatus is equipped with a bottom-entering drive with a high speed impeller system, draught tube and gas sparger. All necessary measuring and control instruments for fermentation were included (pH, P_{O2}, n, p, rH, gas analyses, ...).

For two processes used with different substrates and micro-organisms, the following correlations were evaluated (Seichter, 1987):

- *Torulopsis ethanolitolerans* yeast on ethanol substrate process:

$$k_L a = 0.36 (P_G/V_O)^{0.47} w_S^{0.64}$$

The range of variables:

0.42	$< P_G V_O$	< 6.06	(kW m^{-3})
0.07	$< w_S$	< 0.81	(cm s^{-1})
36	$< \dot{M}_{O2}$	< 278	(mol m^{-3} h^{-1})

The measurements on 50 l and 1 m^3 pilot fermenters (geometrically similar) were included in the treatment of the data.

- *Candida utilis* yeast on waste sulphite liquor fermentation process:

$$k_L a = 0.21\,(P_G/V_O)^{0.44}\, w_S^{\ 0.67} \tag{10}$$

The range of variables:

0.91	$< P_{G/}V_O$	< 5.22	$kW\,m^{-3}$
0.08	$< w_S$	< 0.28	$cm\,s^{-1}$
30	$< M_{O2}$	< 122	$mol\,m^{-3}\,h^{-1}$

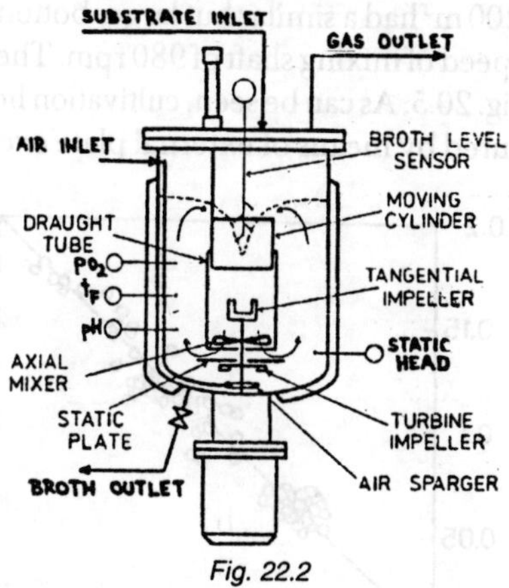

Fig. 22.2

The scatter of data is obvious for both correlations from Fig. 22.3 and Fig. 22.4, Comparing eqn (9) with eqn (10), we can say that the effect of fermntotion growth on the k_L a- value is of great importance.

In the former case, eqn (9) gas-liquid area is probably much larger than in the latter one.

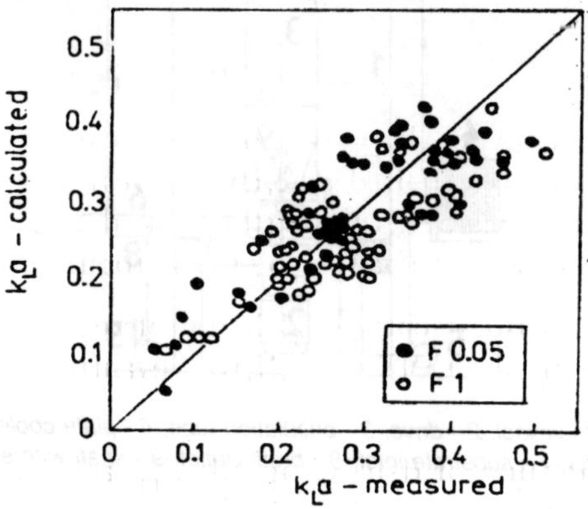

Fig. 22.3

A series of further experiments were carried out during ethanol substrate cultivation of SCP on two other continuous fermenters with 10 m³ and 200 m³, respectively. Mixing systems in both bioreactors were of the same type as laboratory and pilot fermenters (strict geometrical similarity, however, was not observed). Fermenter 10 m³ was equipped with bottom entering drive (17 kW) with mixing shaft rotating on 990 rpm. Fermenter 200 m³ had a similar but larger bottom drive (200 kW) with a rotational speed of mixing shaft of 980 rpm. The latter apparatus can be shown in Fig. 20.5. As can be seen, cultivation heat removal from the broth was secured by means of external plate exchangers.

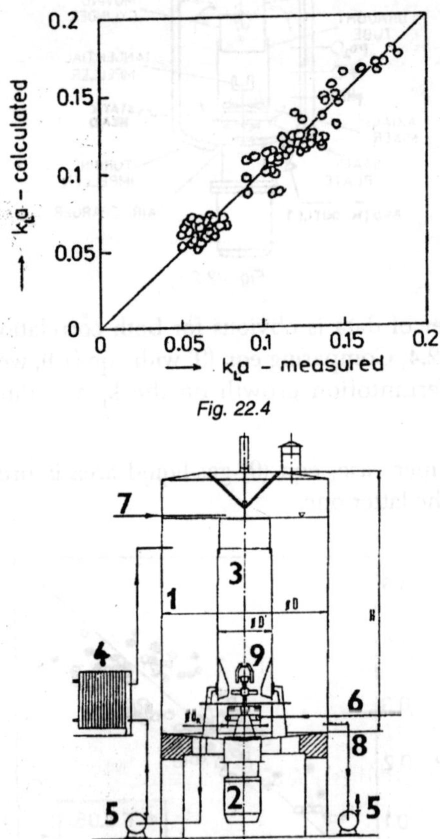

Fig. 22.4

Fig. 22.5.:Key: 1- vessel, 2 - drive, 3 - criculation tube, 4 - plate cooler, 5 - pump 6 - air inlet, 7 - substrate inlet, 8 - broth outlet, 9 - shaft with stirrers.

To make a preliminary evaluation of the measurements on four sizes of fermenters, the modification of correlation (8) was used. Such a treatment of data is shown in Figure 6. It can be indicated that for every size of vessel there is a quite separate solution. Since superficial gas velocity, w_S, increases both with aeration rate, vvm, and with size of the equipment, the following correlation has been proposed.

$$k_La = f[(P_G/V_O), \text{vvm}, w_S] \tag{11}$$

The results of statistical evaluation of such correlation is presented in Fig. 22.7. For fermenters 1, 10 and 200 m³ volumes, eqn (12) has been obtained:

$$k_La = 0.24(P_G/V_O)^{0.38}\ w_S^{-0.41}\ (\text{vvm})^{0.66} \tag{12}$$

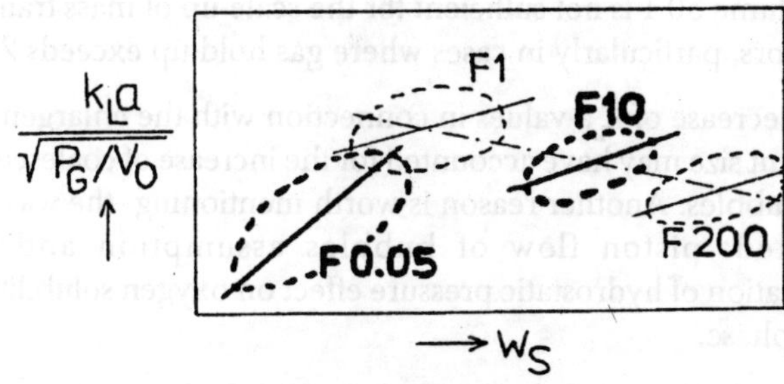

Fig. 22.6

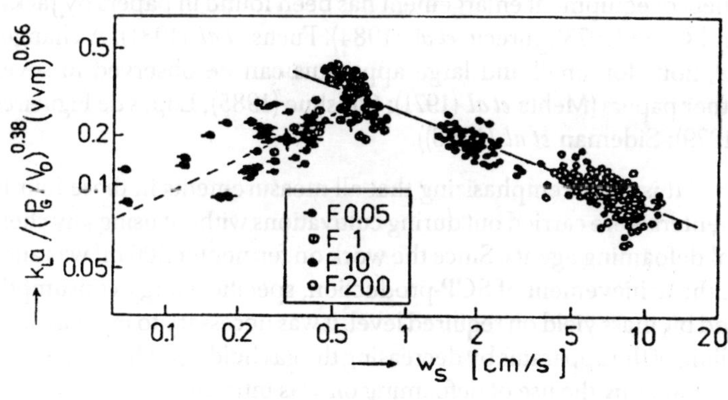

Fig. 22.7

The relationship between w_S and vvm implies the height of liquid filling of a fermenter, H_O:

$$w_S = 1.67\ (\text{vvm})\ H_O \tag{13}$$

This can bring about an alternation of eqn (12) to eqn (14) (Seichter *et al.*, 1990):

$$k_La = 0.2(P_G/V_O)^{0.38} (vvm)^{0.26} H_O^{0.41} \quad (14)$$

The correlations (12) and (14) comprise the ranges of variables:

0.55	<	P_G/V_O	<	4.81	kW m^{-3}
0.36	<	H_O	<	4.4.	m
0.25	<	w_S	<	13.9	cm s^{-1}
0.25	<	vvm	<	2.4	min^{-1}

As can be seen from Fig. 22.7, data evaluated from cultivations performed on the smallest bioreactor 0.05 m^3 volume lie off the scale-up correlation. From this finding, we may state the conclusion that the volume 50 1 is not sufficient for the scale-up of mass transfer in bioreactors, particularly in cases where gas hold-up exceeds 25%.

The decrease of k_La-values in connection with the enlargement of equipment size may have accounted for the increase of coalescence ability of bubbles. Another reason is worth mentioning–the way k_La is evaluated: piston flow of bubbles assumption and the implementation of hydrostatic pressure effect on oxygen solubility in the liquid phase.

The results presented agree with some published ones. A negative effect of equipment enlargement has been found in papers by Jackson and Chen (1978), Jurecic *et al.* (1984). Fuchs *et al.* (1981). A change in relations for small and large apparatus can be observed in several other papers (Mehta *et al.* (1971); Oldshue (1985); Lopes de Figueiredo (1979); Sideman *et al.* (1966)).

It is worth emphasizing that all measurements in these four fermenters were carried out during cultivations without using any chemical defoaming agents. Since the work on fermenter 200 m^3 was aimed at the achievement of SCP-production, specific energy consumption and biomass yield on required level, it was necessary to increase liquid filling of the apparatus by decreasing the gas hold-up. The simplest way to attain this-the use of defoaming oil-was introduced.

Further measurements on F 200 m^3 with ethanol substrate and using defoaming oil, CONTRAMIN, were evaluated by means of eqn (14).

Concentration of defoaming oil, $C_{DF,}$ as a ratio of oil and substrates inflow, changed in the range 0.05 to 10 kg oil/t a.d. SCP.

Evaluated data expressing the influence of C_{DF} on $k_L a$-values can be seen in Fig. 22.8 (every point represents the middle value of approximately a four-hour interval).

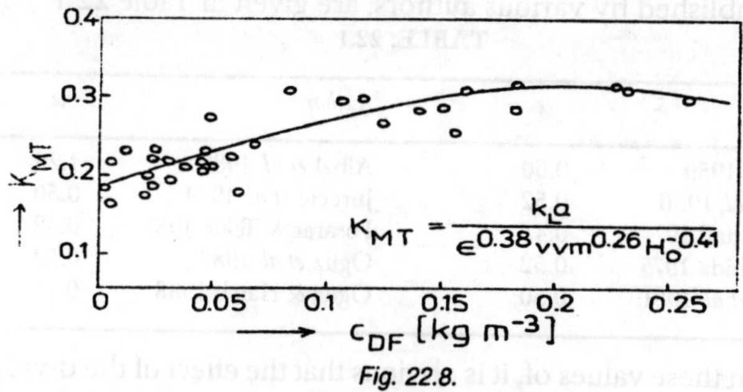

Fig. 22.8.

It is obvious from the curve in Fig. 22.8 that $k_L a$-values can be raised remarkably when the dosing of antifoaming agent is introduced. This surprising phenomenon is valid here only for low concentrations of defoamer. Overdosing leads to rapid drop in interfacial areas and $k_L a$ - value.

Explanation of the course of the dependency in Fig. 8 for ethanol substrate has probably been found in the decrease of liquid phase surface tension.

The effect of defoaming agent concentration on surface tension of this fermentation medium is demonstrated in Fig. 22.9. There is a sharp drop of values with the rise of C_{DF}- concentration up to 0.2 kg CONTRAMIN/m^3.

Further increase of C_{DF} brings about only little change of -value. The above-mentioned value C_{DF} = 0.2 kg m^{-3} in Fig. 22.9 may correspond to the top values of $k_L a$ in Fig. 22.8.

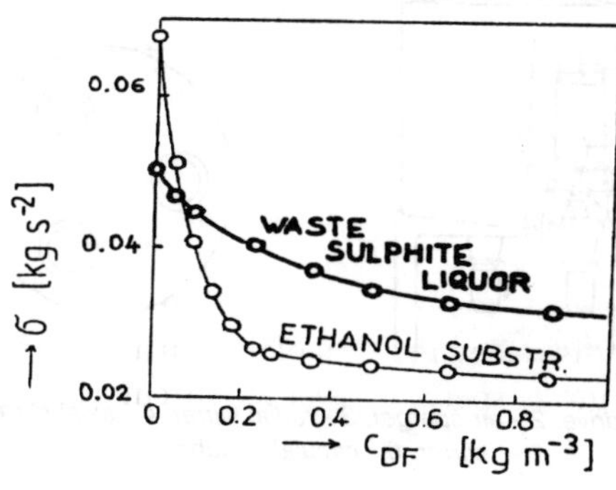

Fig. 22.9.

Many papers have dealt with the influence of surface tension on k_La-values. Exponents of the relation:

$K_L a \sim \sigma \alpha$ (15)

as published by various authors, are given in Table 22.1.

TABLE: 22.1

Author	*α*	*Author*	*α*
Calderbank 1959	0.60	Albal *et al.* 1983	1.29
Sideman *et al,* 1966	0.52	Jurecic *et al.* 1984	0.50
Perez & Sandall 1974	0.45	Posarac & Tekic 1987	0.62
Yagi & Yoshida 1975	0.52	Oguz *et al.* 1987	0.52
Nishikawa *et al.* 1981	0.50	Oguz & Hatch 1988	0.13

From these values of, it is obvious that the effect of the σ-value is relevant and is in plausible agreement with our results obtained on industrial fermenter. This conclusion, however, cannot be generalized for either arbitrary defoamers or fermentation broths.

Acquired experience together with the discussed scale-up procedure for oxygen transfer rate in high-speed impellers bioreactor were utilized in the design of large scale fermenters 800 m^3 in volume for SCP production from saccharides contained in waste sulphite liquors. The apparatus is shown schematically in Fig. 22.10. As can be seen, this is an application of three drives with 200 m^3 fermenter) entering the bottom of one apparatus 800 m^3 fermenter) entering the bottom of one apparatus 800 m^3 in volume. They all work independently of each other. In order to achieve the required process parameters, it was necessary to apply various defoaming agents (for instance: type STRUCTOL : 15 kg/t a.d. SCP; type CONTRAMIN: 4.5 kg/t a.d. SCP.)

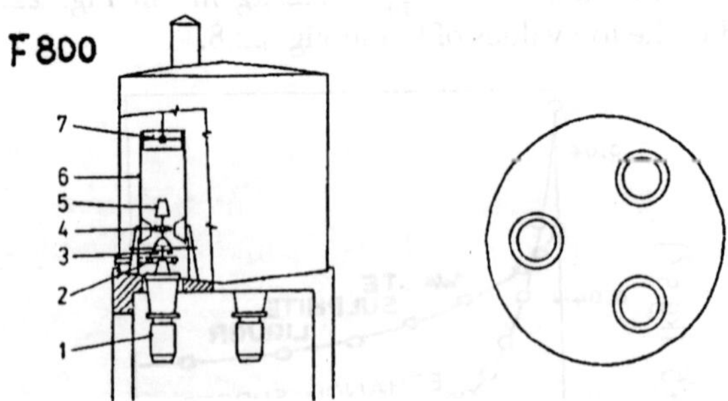

Fig. 22.10: key: 1 - drive, 2 - air sparger, 3 - radial stirrer, 4 - axial stirrer, 5 - tangential stirrer, 6 - circulation tube.

Oxygen transfer measurements in fermenters 800 m^3 have proved good agreement in k_La-values with scale-up correlation (12), indicated in Fig. 20.11. This seems to be rather a coincidence due to at least two factors:

- negative effect of a 'worse' substrate (sulphite liquors)
- positive effect of defoaming agents dosing.

Nevertheless, the difference between both industrial fermenters 200 m^3 and 800 m^3 as regards kLa- values is not significant.

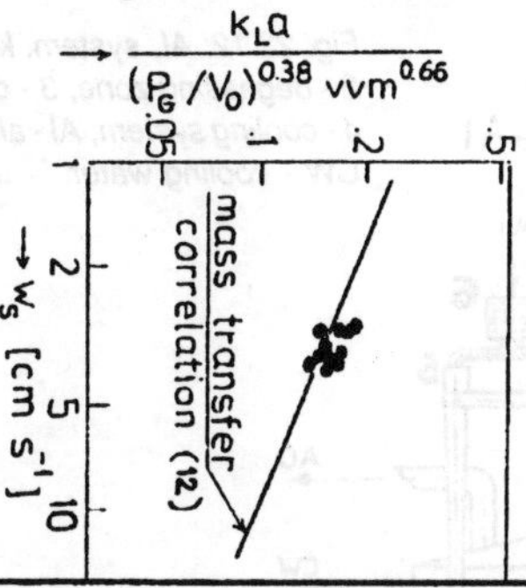

Fig. 22.11.

Production of SCP on a large scale has been realized, in other types of bioreactors of some parameters of the former with the following systems:

- air-lift (AL), see Fig. 22.12 (Westlake, 1986; Hines, 1977)
- deep jet (DJ), see Fig. 22.13 (Vogelbusch) is presented in Table 22.2. Some advantages of mechanically mixed (MM) bioreactors appear obvious.

Data in Table 22.2 are rather influenced by differences in substrates: especially with their distinct specific oxygen consumption for production of SCP. Nevertheless, the highest values of oxygen transfer rates and productivity are, however, with the ICI air-lift type. The other way round, this advantage is accompanied by higher specific energy consumption Ax. From the latter point of view, MW and DJ systems seem to be on the same level. Al and DJ systems, however, are heavier and probably more costly.

Data obtained for working bioreactors of various types both by means of direct measurement and from the literature have been used for k_La-values estimates. The following correlation have been proposed for the evaluation of industrial bioreactor performance (SCP and baker's yeast production):

$$K_L a \approx c\ (\Sigma\ P_f/V_O)^{0.5}\ H_O^{\ \beta}$$

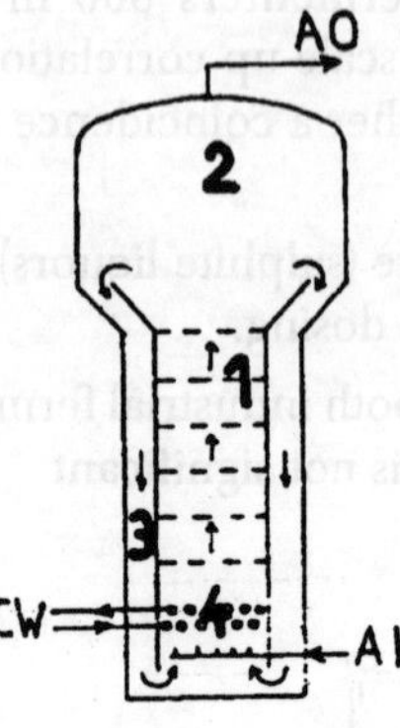

Fig. 22.12: AL system. key: 1 - riser section, 2 - degassing zone, 3 - downcoming section, 4 - cooling system, AI - air inlet, AO - air outlet, CW - cooling water.

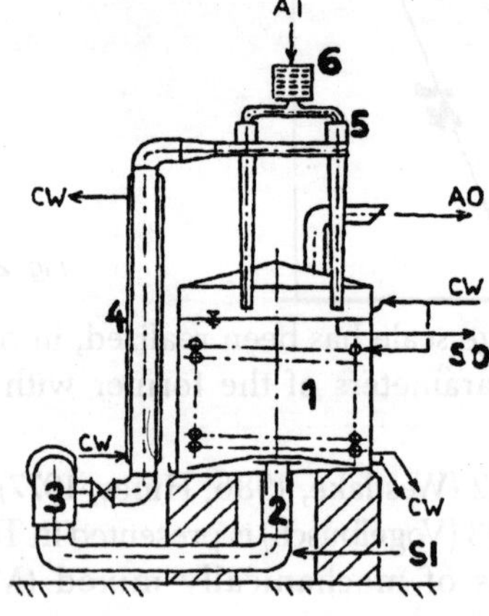

Fig. 22.13: DJ system. Key: 1 - vessel, 2 - circulation tube, 3 - circulation pump, 4 - tube cooler, 5- ejector, 6- air filter, AI, AO, CW as before, SI substrate inlet, SO broth outlet.

TABLE: 22.2.
VARIOUS AEROBIC BIOREACTOR PARAMETERS (PRODUCTION OF SCP). METH., METHANOL; WS, WASTE SULPHITE LIQUOR; N-AL., N-ALCANS; ETH., ETHANAOL.

Type *Country*	*AL* *ICI-GB*	*DJ* *IZ-Germany*		*MM* *ZVU-Czech*	
Substrate:	Method	ws	n-al	ws	eth.
$V_O(m^3)$	1050	220	1000	280	100
$M_{O2}(mol\ m^{-3}h^{-1})$	320	115	215	120	180
$pr_{vo}(kg\ m^{-3}h^{-1})$	6.0	3.5	3.3	4.2	4.1
Ax $(kWh\ kg^{-1})$	1.12	0.83	1.94	0.83	0.86
Weight (t)	600	95	270	82	30

Total energy dissipated in a unit of time in a unit of volume of a bioreactor is the sum of specific power inputs entering fermentation broth:

$$\sum P_F = P_G + P_{IS} + P_H \qquad (17)$$

where

$$P_{IS} = \dot{V}_G\,[\rho_G\,(W_D{}^2/2) + P_{AT}\ \text{in}\ (p_D/p_2)] \tag{18}$$

$$PH = VL\,\rho L\,g\,HF$$

Results of the data treatment are presented in Fig. 22.14. Despite the variety of processes, substrates, micro-organisms, technical level and the type of equipment, the correlation plausibly covers a wide scope of conditions. The effect of liquid filling height is less pronounced here than in eqn (14) . It is worth remembering the way in which the k_La-values have been determined here, using eqn (5): this could contribute to generalizing all production plant data.

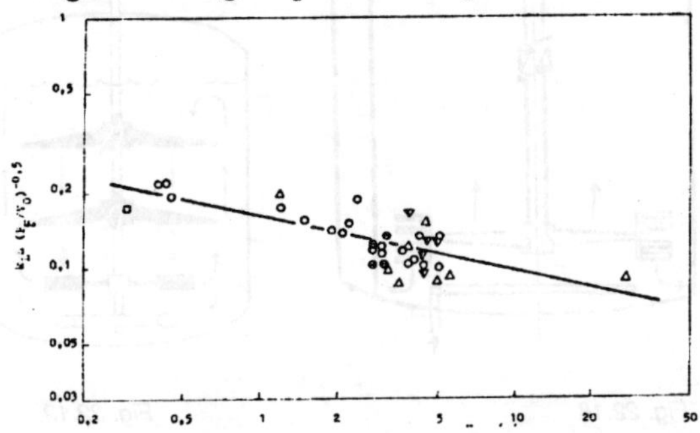

Fig. 22.14: Key: O- mechanically mixed, θ - with sell-inducing impellers,▼- deep jet system, Δ -air lift, ▯- air column.

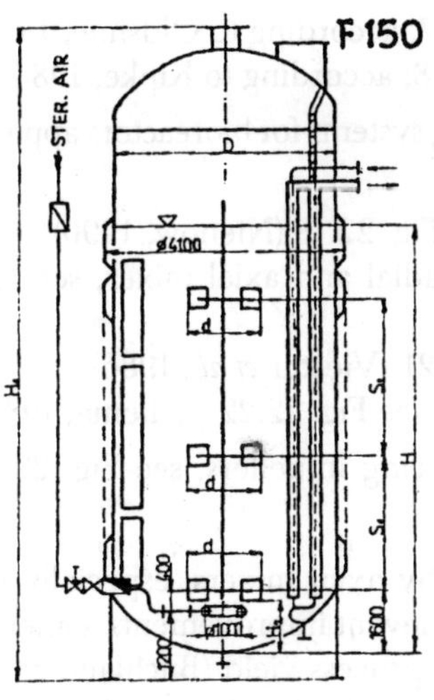

Fig. 22.15

It would be convenient to recall, however, more typical representatives of agitated bioreactors. The most common type, equipped with turbine stirrers, is shown in Fig. 22.15. This version was used on our 150 m³ fermenter for production of citric acid and have been widely applied for many aerobic fermentations during the last 40 years. Recently, however, several new mixing systems have been successfully tested. Mostly as a multiple impellers lay-out, the following mixers have been designed or retrofitted:

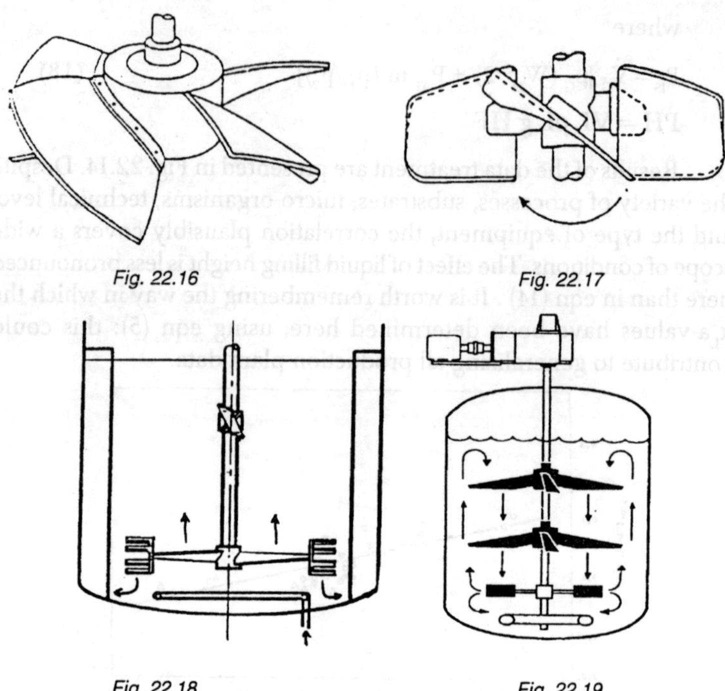
Fig. 22.16
Fig. 22.17
Fig. 22.18
Fig. 22.19

Prochem Hydrofoil, see Fig. 22.16, according to Buckland *et al.*, 1988

- Lightnin A 315, see Fig. 22.17, according to Oldshue, 1990
- Ekato Intermig, see Fig. 22.18, according to Kipke, 1984.

An alternative solution of mixing systems for bioreactors appears to be the combined agitator systems:

- axial and radial mixers, see Fig. 22.18 (Nienow, 1990)
- draught tube system with radial and axial mixer, see Fig. 22.21 (Anderson *et al.*, 1982)
- Waldhof system, see Fig. 22.21 (Viestur *et al.*, 1986)
- counterflowing axial mixers, see Fig. 22.22 (Achema, 1985).

For some applications self-inducing impellers, see Fig. 22.23 (Viestur *et al.*, 1986), are popular too.

Substitution of turbine stirrers by axial mixers, especially for viscous broths, has proved to bring relevant improvements, either in saving energy or with an increase in process yield (Buchland *et al.*, 1988; Kipke, 1985). This may be due to both sufficient dispersion ability to axial mixers and enriched blending capacity of the mixers.

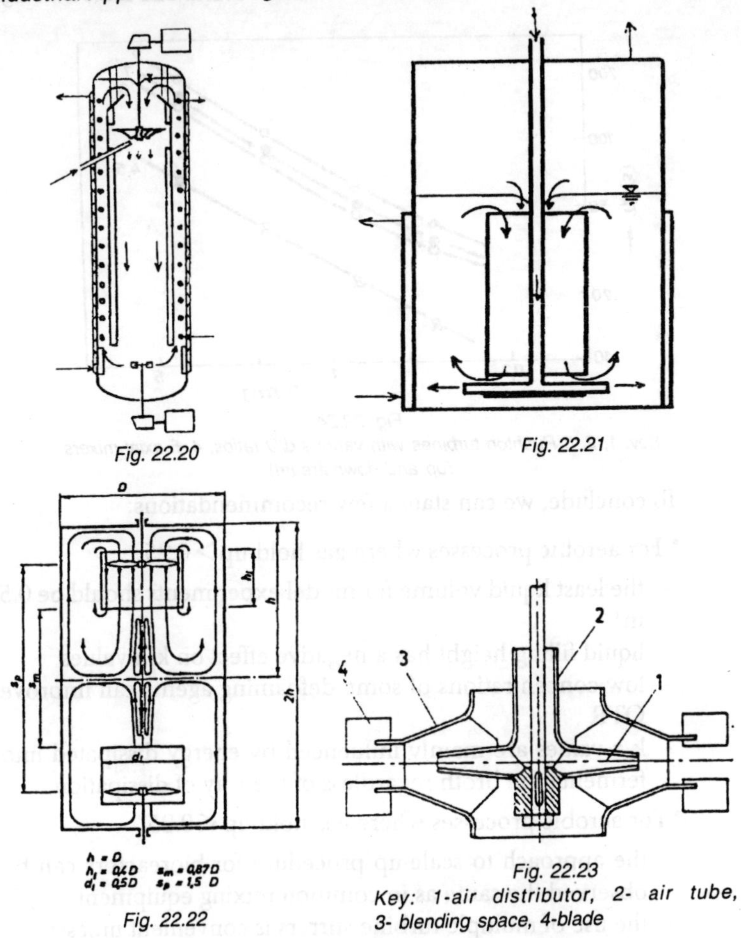

Fig. 22.20

Fig. 22.21

Fig. 22.22

Fig. 22.23

Key: 1-air distributor, 2- air tube, 3- blending space, 4-blade

An example of the comparison of blending times for two arrangements of multiple mixers on one shaft (Seichter *et al.*, 1990):

- 5 axial mixers (d/D = 0.625)
- 3 turbine stirrers (d/D = 0.36, 0.40, 0.44)

can be seen in Fig. 22.24. For the same level of specific power inputs (P/V_O) with both arrangements, blending times for axial mixers are much shorter than for turbine stirrers irrespective of the size of equipment.

Kipke (1984) presented another comparison of both systems– from the point of shear stresses, for constant (P/V_O)-values, shear stresses developed by turbine stirrer were 7-times greater than those developed by axial mixer (Intermig).

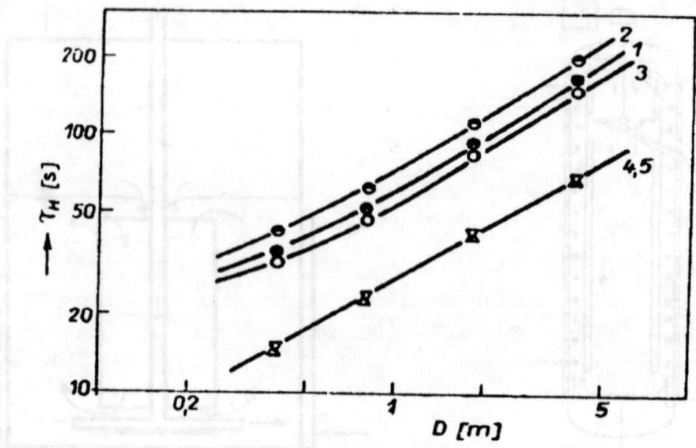

Fig. 22.24

Key: 1, 2, 3 Rushton turbines with various d/D ratios, 4, 5 axial mixers (up and downstream)

To conclude, we can state a few recommendations:

* For aerobic processes where gas hold-up > 0.25:

- the least liquid volume for model experiments should be 0.5 m^3
- liquid filling height has a negative effect on $k_L a$-values
- low concentrations of some defoaming agents can improve OTR
- $k_L a$-values are mainly influenced by energy dissipated into fermentation broth regardless of the way of dissipation.

* For aerobic processes where gas hold-up < 0.25:

- the approach to scale-up procedure for bioreactors can be observed the same as in common mixing equipment
- the use of multiple turbine stirrers is convenient unless:
  fermentation broth is viscous
  micro-organisms are shear stress sensitive
  rapid blending of broth is required
- several types of axial mixers can successfully replace radial ones.

REFERENCES

ACHEMA. 1985. Information leaflet. TU Berlin, Inst. Chemieingenieurtechnik, Achema, Frankfurt a.M.

Albal, R.S., Shah, Y.T., Schumpe, A. and Carr, N.L. 1983. *Chem. Eng.* J, 27, 61-80.

Anderson, C., Le Grys, G.A. and Solomons, G.L. 1982. *The Chem. Eng.*, p. 43.

Buckland, B.C., Gbewonyo K., Di Masi, D., Hunt, G., Westerfield, G. and Nienow, A.W. 1988. *Biotechnol. Bioeng.*, 31, 737.

Calderbank, P.H. 1959. *Trans. Instn. Chem. Eng.*, 37, 73.

Fuchs, R., Ryu, D.D.Y. and Humphrey, A.E. 1971. *IEC Proc. Des. Develop.*, 10, 190-96.

Henzler, H.J. 1982. Chem. Ing. Tech., 54,461-76.

Hines, D.A. June 1977. Proc. Ist Europ. Congr. Biotechnol., Interlaken, Switzerland. pp. 55-64.

Jackson, M.L. and Chen, Ch. Ch. 1978. *AICHE J.*, 24, 63-71.

Jurecic, R., Berovic, M., Steiner, W. and Koloini, T. 1984. *Canad. J. Chem. Eng.*, 62, 334-39.

Kipke, K. 1984. *Chemie-Technik,* 13, 46.

Kipke, K. 1985. EKATO Information. Mixing Technology Reports, Brochure for EKATO.

Kossen, N.W.F. and Oosterhuis, N.M.G. 1985. Modelling and scaling-up of bioreactors. In *Biotechnology.* VCH Verlagsgesellschaft, Weinheim. Vol. 2, Ch. 24.

Lopes de Figueiredo, M.M. and Calderbank, P.H. 1979. *Chem. Eng.* Sci., 34, 1333-38.

Mehta, V.D. and Sharman, M.M. 1971. *Chem. Eng. Sci.*, 26, 461

Nienow, A.W. 1990. *Chem. Eng. Prog.,* 61-71.

Nishikawa, M., Nakamura, M. and Hashimoto, K. 1981. J. Chem. Eng. Jpn, 14, 227-32.

Oguz, H., Brehm, A. and Deckwer, W.D. 1987. Chem. Eng. Sci., 42, 1815-22.

Ogut, A. and Hatch, R.T. 1988. *Can. J. Chem. Eng.*, 66, 79-85.

Oldshue, J.Y. 1985. Biotech. Adv., 3, 219-37.

Oldshue, J.Y. 1990 . The comparison between axial flow and radial flow impellers for gas-liquid mass transfer. Paper presented at 10th CHISA Congress, Prague, Czechoslovakia.

Perez, J.F. and Sandall, O.C. 1974. *AICHE J.*, 20, 770-75.

Posarac., D. and Tekic, M.N. 1987. *AICHE J.*, 33, 497-99.

Seichter, P., *Coll. Czech. Chem. Commun.*, 52, 2654-63.

Seichter, P., Kvasnicka, J., Mazoch, J. and Medek, J. 1990. *Chem. prumysl.*, 40/65, 343-46.

Seichter, P., Pesl, L. and Vranek, M. 1990. Performance of large scale bioreactors. 10th Int. Congr. , CHISA. Prague, Czechoslovakia.

Sideman, S., Hortacsu, O. and Fulton, J.W. 1966. *Ind. Eng. Chem.* 58, 32.

Van't Riet, K. 19799. *IEC Proc. Des. Develop.*, 18, 367-75.

Viestur, U.E., Kuznecov, A.M. and Savenkov, V.V. 1986. *Sistemy Fermentacii.* Vyd. Zinatne, Riga, Latvia.

Vogelbusch, brochure.

Westlake R. 1986. *Chem. Ing. Tech.*, 58, 934-37.

Yagi, H. and Yoshida, F. 1975. *IEC Proc. Des. Develop.*, 14, 488-93.

23

DOWNSTREAM PROCESSING

I. ROUSAR
Czech Technical University, Prague, Czech Republic

Introduction

Isolation and recovery of the fermentation product are as important and sometimes more expensive than the fermentation itself. Whereas the price of bulk bioproducts is decreasing with the amount of world production (Hacking, 1986), the selling price of various biological specialities is inversely proportional to their concentration in starting material (Dwyer, 1984), see Fig. 23.1.

Developments in genetic engineering, the production of monoclonal antibodies from hybridoma cells and systems for continuous recycle fermentation introduced the need for new recovery and separation processes.

Besides improvements in the design and operation of traditional separations such as centrifugation, filtration, absorption, ion exchange, and even sedimentation, recent advances in relatively new processes such as membrane separations, chromatography, and electrophoretic separation have expanded the option enormously (Bungay and Belfort, 1987).

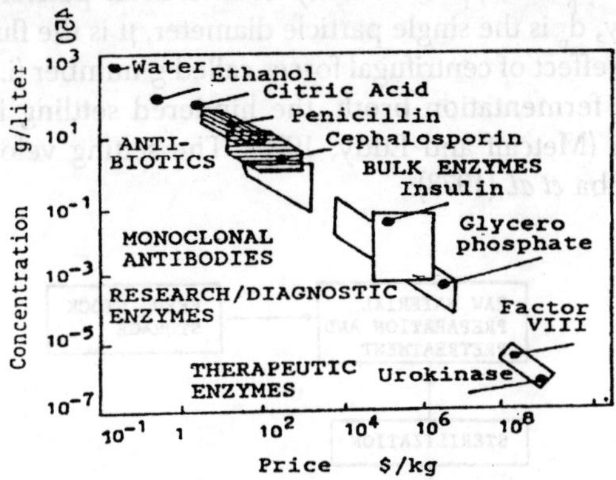

Fig. 23.1: Concentration in starting material of various substances and their selling price (from Dwyer, 1984).

THE GENERAL SCHEME OF DOWNSTREAM PROCESSING

Every bioprocess consists of three main parts: upstream, fermentation and downstream, see Fig. 23.2.

The termentation broth after fermentation contains cells and liquid medium with the extra cellular products. The first step is cell separation or cell harvesting. If the intracellular products are to be separated the cell disintegration step has to precede the isolation and purification. Every step can be achieved by different unit operations as depicted in the right side of Fig. 23.2.

BIOMASS SEPARATION

Biomass was traditionally separated from the liquid by differential migration processes (sedimentation, centrifugation and filtration) and flotation. At present the most attention is paid to the membrane process–microfiltration.

The design of a *sedimentation* and *centrifugation* unit starts from the terminal velocity U_{og} of motion of one particle in gravitational field:

$$U_{og} = \frac{g d_p^2 (\rho_p - \rho_f)}{18\mu} \quad (1)$$

and in centrifugation field

$$U_{oc} = \frac{g Z_p^2 (\rho_p - \rho_f)}{18\mu} \quad (2)$$

where p_p and p_f are density of microbial particle and fluid respectively, d_p is the single particle diameter, μ is the fluid viscosity and z is the effect of centrifugal forces called g-number (i.e., $Z = r\,\omega^2/g$). For the fermentation broth, the hindered settling is the usual mechanism (Metcalf and Eddy, 1979). The setting velocity is then given by Aiba *et al.* (1973):

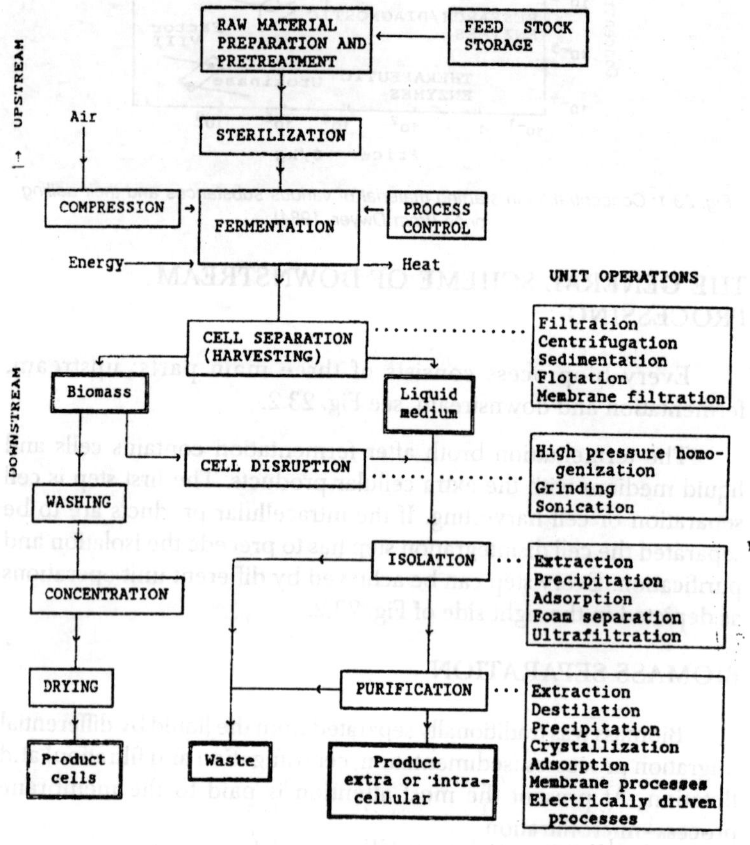

Fig. 23.2: The general scheme of bioprocess

$$\frac{U}{U_{og}} = \frac{1}{1 + \alpha' C_v^{B}} \qquad (3)$$

Where $\alpha' = 1 + a\,c_v^{B}$ with A=305, B=2.84, $0.15 < c_v, < 0.5$ for irregular particles, and A= 229, B = 3.34, $0.2 < c_v < 0.5$ for spherical particles. For dilute suspension where volumetric ratio of solid particles c_v is less than 0.15, A= 1 to 2.

Hazen (1904) realized that settling in an inclined tube is faster than in a vertical one. Acrivos and Herbolzheimer (1979) analyzed inclined batch settler and inclined continuous flow settlers and found the optimum unclination from vertical axes to be 60°C. This phenomena is utilized in lamellar settlers and disc centrifuges.

The settling can be enhanced by adding flocculents or electrolytes which destabilize the electrostatic double layer. The cells tend to behave as colloids usually charged with negative charge and therefore the particles are kept in a stable position by the electrostatic repulsive force (Belfort, 1987).

Examples of industrial devices for sedimentation and centrifugation are presented in Figs. 23.3 and 23.4.

Filtration can be performed in two different modes. In deep bed filtration the particles are caught inside the porous medium, while in cake filtration the cake itself represents the filtration media. The major problem associated with filtration of fermentation broth is the compressibility of the accumulated solids on the filter surface and the resultant decrease of permeability during the filtration cycle. Diatomaceous earth and pearlite from volcanic rock are used as the filtration aid to inhibit cake compression. The filtration rate comes from the modified Poiseuille relationship:

$$\frac{\dot{V}}{S} = \frac{\Delta p}{\mu\,[\alpha\,(W/S) + r]} \qquad (4)$$

where $\dot{V}$ is volumetric flow rate, S is area of filtration surface, μ is broth viscosity, α is average specific cake resistance, W is mass of accumulated biomass, and r is resistance of filtrate medium. Typical filtration rates for a variety of applications are given in Table 23.1.

TABLE: 23.1
TYPICAL FILTRATION RATES (MOO-YANG *ET AL.*, 1985).

Product	*Organism*	*Filter type*	*Filtration rate (m^3/m^2 h)*
Kanamycin	*S. kanamyceticus*	Vacuum precoat	0.00074
Lincomycin	*S. lincolnensis*	Vacuum precoat	0.0033
Neomycin	*S. fradiae*	Vacuum precoat	0.0011
Penicillin	*Unknown*	Vacuum precoat	0.012-0.016
Protease	*B. subtilis*	Vacuum precoat	0.0009-0.0037
Disrupted yeast cake	*Unknown*	Rotary vacuum	0.0026

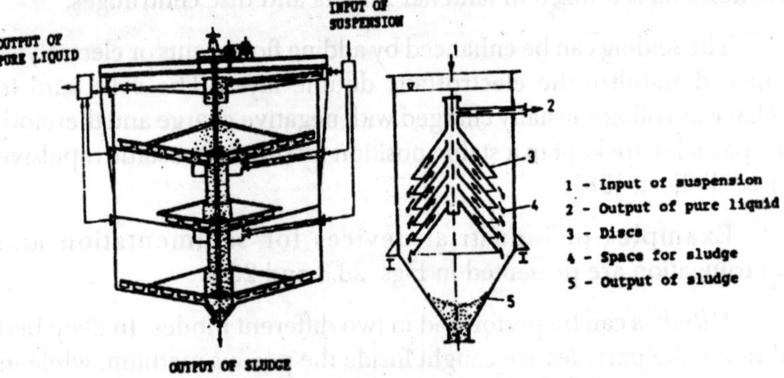

Fig. 23.3: Two types of settlers (Novák et al ., 1989)

Although filtration has been the major primary tool for antibiotics in the past, this unit operation is slowly being replaced by other approaches such as centrifugation and the ion exchange.

The continuous rotary vacuum filters are usually the choice for the large scale filtration of fermentation broth, see Fig. 23.5. As drum rotates during the filtration cycle, a thin layer of biomass and filter aid accumulates on the surface of the precoat and builds up during the filtration cycle. Upon further dewatering followed by washing of the formed cake, a doctor blade removes a thin layer of the cake and filter aid and expose a fresh surface for the next cycle.

The other fast expanding technique for the cell separation, microfiltration, is covered later under membrane processes.

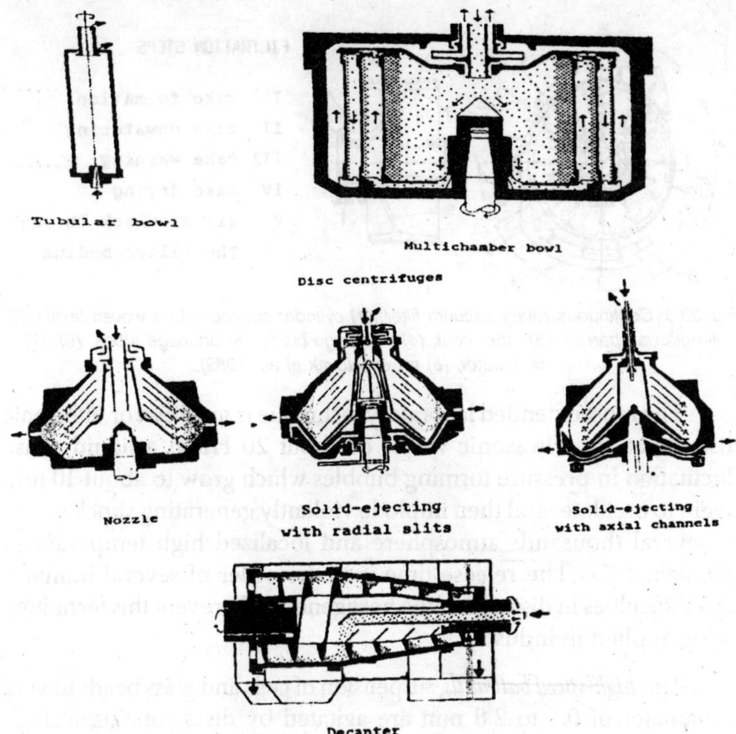

Fig. 23.4: Centrifuges used in downstream processing (Axelsson, 1985).

CELL DISRUPTION

The content of microbial cell is protected from rupture by a strong, rigid, outer cell wall, see Table 23.2.

With currently available methods, such as high-speed ball mills and high-pressure homogenization, disruption is a high-cost, energy intensive unit operation (Mogren *et al.*, 1974).

TABLE 23.2
SHEAR STRESS CAUSING BREAKAGE OF THE CELL WALL (MERSMANN *ET AL.*, 1990).

Type of cells	*Shear stress causing breakage*
Bacteria	10^7 - 10^8 Pa
Yeast, moulds	10^7 Pa
Plant cells	10^6 Pa
Animal cells	0.005 - 500 Pa

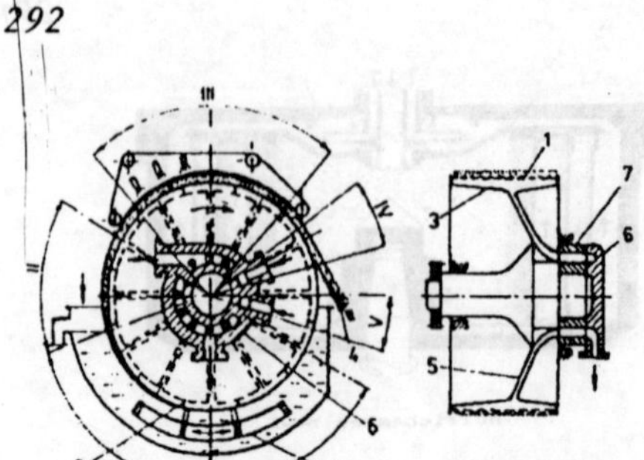

Fig. 23.5: Continuous rotary vacuum filter: (1) cylinder surface with a woven textile, (2) lengthwise barrier, (3) inner coat, (4) discharge blade, (5) drainage tubes, (6), (7) distributor, (8) mixer (Novak et al., 1989).

The most extended laboratory technique is *sonication* or ultrasonic disintegration. Ultrasonic waves of about 20 Hz in a liquid cause fluctuation in pressure forming bubbles which grow to about 10 μm, begin to oscillate, and then implode violently generating shock waves of several thousands atmosphere and localized high temperatures (Doulah, 1977) The release time is in the order of several minutes. The difficulties in dissipating the heat generated prevent this technique being applied in industry.

The *high-speed ball mills*, suspension of cells and glass beads having a diameter of 0.1 to 2.8 mm are agitated by discs rotating at high speed. The peripherical speed of the stirrer ranges from 9 to 22 m/s and the time for 10% disintegration is in the order of minutes (Aiba *et al.*, 1973). The disruption of cells is generally a first-order process. The rate of disintegration is proportional to the cell concentration. The energy consumption ranges from 0.05-0.56 kWh/kg of dry cells (Engler, 1985).

Other industrial cell disintegration is *high pressure homogenization*. Disruption is accomplished by passing cell suspension under high pressure of 55 MPa through an adjustable, restricted orifice discharge valve. The degree of disintegration in one pass through the valve lies between 10 and 60%. The energy consumption ranges from 0.17 to 0.33 kWh/kg (Engler, 1985).

PRODUCT RECOVERY FROM LIQUID MEDIUM

When a desired product is soluble in organic solvent, *extraction* is the best option to achieve a rough separation. The recovery of a large amount of penicillin during the Second World War was accomplished by extraction. In the first step the penicillin is extracted from fermentation broth by the organic solvent at low pH of 2 at which the penicillin prefers solvent. In the second step using the alkali solution the pH is changed to 7.5 and this causes penicillin to be dissolved back into a pure water. The concentration of penicillin in the final solution can be 100 times greater than in the feed (Hatton, 1985).

The most common extractor is of Podbielniak type which is depicted in Fig. 23.6. This extractor can handle up to 1 m^3 of fermentation broth per minute.

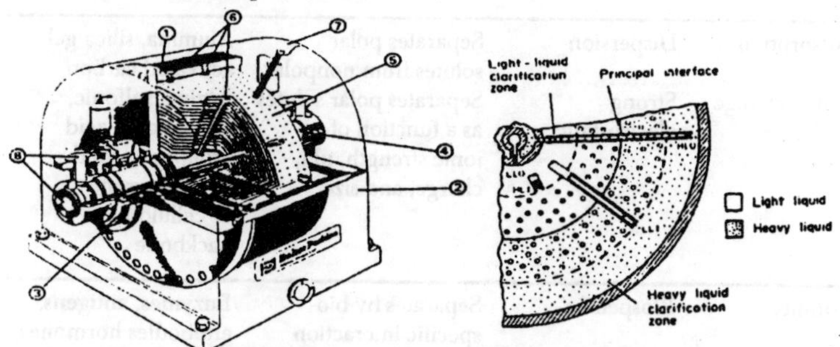

Fig. 23.6.: Podbielniak multistage counter-current extractor: (1) rotor and shaft assembly, (2) base, (3) bearings, (4) multi-V-belt drive, (5) concentrically perforated elements, (6) liquid distribution tubes, (7) 'Asco' tubes to permit access for cleaning and flushing, (8) inlet and outlet ports (Moo-Yang et al., 1985)

The liquid-liquid extraction has been used also for the recovery of biopolymer like proteins or nucleic acid despite the fact that these biopolymers are mostly insoluble in organic solvents. In this case the two phase system is created by the addition of hydrophilic polymers into an aqueous solution. Several processes for the isolation of enzymes have been developed and are operated on the pilot plant (Kula *et al.*, 1982) and industrial scale.

Sorption techniques are used to separate solutes from fermentation broth. They can be performed in a batch mode, fixed bed column or fluidized bed. Batch techniques are generally used more commonly than fixed bed column, which can run in a chromatography mode (injection of sample to carrier) or in cycles of absorption and desorption. Every solution is a mixture of solutes with different sorption affinities. Only chromatography can take from the mixture components whole sorption affinity lies somewhere between the absorption affinities of other solutes. Batch techniques and fixed bed in cyclic mode can divide the mixture of solutes with different absorption affinities into two streams. One contains the solute with sorption affinity lower than critical affinity. The components with higher affinity than critical are concentrated in the second stream. The value of critical affinity depends on the operating parameters of process mainly on mean residence time.

Different physico-chemical mechanisms can cause the sorption of solutes on porous sorbent, see Table 23.3.

TABLE: 23.3

SORPTION MECHANISMS (BELFORT, 1987).

Mechanism	*Dominant Interaction*	*Comments*	*Active surface*
Adsorption	Dispersion	Separates polar solutes from nonpolar	Alumina, silica gel activated carbon
Ion exchange	Strong electrostatic	Separates polar solutes as a function of ionic strength, pH, charge, and size	Amine, sulfonic, phosphoric, acid groups attached to polystyrene or cellulose backbone
Affinity	Biospecific	Separates by bio-specific interaction	Enzymes, antigens, antibodies hormones, attached or immobilized to silica or other carriers
Gel filtration	Hydrophilic plus low electro-static	Uses hydrophilic gels with weak anion and cation exchanges to separate polar from nonpolar according to their size	Dextrane, agarose, polyacrylamide
Liquid-liquid partition	Polar versus nonpolar	Partition coefficient	Polyethylene glycol and dextran or certain aqeous solution

Absorption chromatography has been used in the initial recovery of the high-cost extracellular enzymes on a large scale. The scale up procedure of chromatography column is clearly described by Ladish (1987). Melling and Phillips (1985) give an excellent review of the basic principles of ion exchange procedure, the properties of the materials used and their application to enzyme purification. Weetall (1979) gives an extensive review of some of the molecules that have been purified by affinity chromatography while Yarmush et al. (1985) gives deep insight on gel filtration sometimes called molecular sieve chromatography.

Electrically driven processes are based on the differences in electrophoretic mobility of a solute. Besides electrodialysis (Jain and Reed, 1985), which has been used for large-scale desalting applications (>4000 m^3/m^2/day), the other processes like electrofiltration, electrophoresis, isoelectric focusing, forced flow electrophoresis, electrodecantation, and isotachophoresis are not much larger than analytical and some preparative scale.

MEMBRANE PROCESSES

Membrane processes are now the fastest spreading large scale separations. The membrane processes have been used to remove ionic molecules (hyperfiltration HP or reverse osmosis NO), macromolecules (ultrafiltration UF) and microparticles like cells (microfiltration MF) (Moo-Yang *et al.*, 1985) and for other applications, see Table 23.4.

TABLE 23.4: MEMBRANE PROCESSES.

Name	Schema	Driving force	Permeate / Retenate	Practical example
Reverse osmosis (Hyper-filtration)	 L → L, L	$\Delta p-\sigma\Delta\Pi$	Solvent / Small molecules 10 – 50 MW	Seawater desalination
Ultra-filtration	 L → L, L	Δp	Solvent + small molecules / Macromolecules 500 – 300000 MW	Separation of macromolecules
Micro-filtration	 L → L, L	Δp	Solvent + small molecules macromolecules / Microparticle	Cell separation
Dialysis	 L, L → L, L	Δc $\Delta p=0$	Small molecule + macromolecule / Small molecule + macromolecule	Artificial kidneys
Electro-dialysis	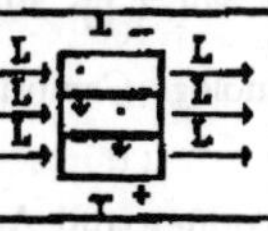 L, L, L → L, L, L	ΔE	Anions / Cations	Separation of citric acid from fruit juice
Pervaporation	 L → V, L (vacuum)	Δc	Vapor of one component / Liquid	Separation of azeotropes

small ionic molecules • macromolecules o microparticles

Synthetic membranes are used to 'filter' the particles, macromolecules, or ions. The retained species are concentrated on the up-stream side of the membrane and cause the reduction of the permeation flux owing to (1) increased osmotic back pressure, (2) hydraulic resistance of the gel or cake layer and (3) adsorption onto the membrane surface. These phenomena, called gel or concentration polarization, are now the limiting factors in the performance of most of the industrial applications. An excellent review of theory, materials, membrane modules and membrane application in downstream processing has been given by Belfort (1987).

The downstream processing still represents the huge challenges for the further research on new separation techniques, optimization and scale up. In this article, some of the major separation techniques used in downstream processing have been mentioned as they were dealt with during the UNESCO Course.

REFERENCES

Acrivos, A. and Herbolzheimer, E. 1979. *J. Fluid Mech.*, 92, 435.

Aiba, S., Humphrey, A.E. and Mills, N.F. 1973. *Biochemical Engineering*. Academic Press, New York. 2nd edn.

Axelsson, H.A.C. 1985. In M. Moo-Yang, C.L. Cooney and A.E. Humprey (eds), *Comprehensive Biotechnology*. Pergamon Press. Vol. 2, pp. 325-46.

Belfort, G. 1987. In H.R. Bungay and G. Belfort (eds), *Advanced Biochemical Engineering*, John Wiley & Sons, New York. pp. 219-38.

Bungay, H.R. and Belfort, G. (eds) 1987. *Advanced Biochemical Engineering*. John Wiley & Sons, New York.

Doulah, M.S. 1977. Mechanism of disintegration of biological cells in ultrasonic cavitation. *Biotechnol. Bioeng.*, 19, 649-60.

Dwyer, J.L. 1984. *Biotechnology*, 2 (11) 957-64.

Engler, C.R. 1985. In M. Moo-Yang, C.L. Cooney and A.E. Humprey (eds), *Comprehensive Biotechnology*. Pergamon Press. Vol. 2, pp. 305-24.

Hacking, A.J. 1986. Economic Aspects of Biotechnology. Cambridge University Press, London.

Hatton, T.A. 1985. In M. Moo-Yang, C.L. Cooney and A.E. Humprey (eds), *Comprehensive Biotechnology*, Pergamon Press. Vol. 2, pp. 439-50.

Hazen, A. 1904. *Trans. ASCE*, 53, 45.

Jain, S.M. and Reed, P.B. 1985. In M. Moo Yang, C.L. Cooney and A.E. Humprey (eds), *Comprehensive Biotechnology*. Pergamon Press. Vol. 2, pp. 575-90.

Kula, M. R., K.H. and Hustedt, H. 1982. Purification of enzymes by liquid-liquid extraction. *Adv. Biochem. Eng.*, 24, 73-118.

Ladish, M.R. 1987. In H.R. Bungay and G. Belfort (eds), *Advanced Biochemical Engineering.* John Wiley & Sons, New York. pp. 79-102.

Melling, J. and phillips, B.W. 1985. In a Wiseman (ed), *Handbook on Enzyme Biotechnology.* Ellis Harwood, Chichester. pp. 55-88, 181-202.

Mersmann, A., Schneider, G., Voit, H. and Wenzig, E. 1990. Chem. Eng. Technol., 13, 357- 70.

Metcalf and Eddy. 1979. *Waste-Water Engineering, Treatment, Disposal, Reuse.* McGraw-Hill, New York.

Mogren, H., Lindblom, M. and Hedesgong, G. 1974. Mechanical disintegration of micro-organisms in an industrial homogenizer. *Biotechnol. Bioeng.*, 16, 261-74.

Moo-Yang, M., Cooney, C.L. and Humprcy A.E. (eds), 1985. *Comprehensive Biotechnology.* Pergamon Press.

Novak, V., Rieger, F. and Vavro, K. 1989. *Hydraulic Unit Operations.* SNTL, Prague. In Czech.

Weetall, H.H. 1979. In E.S. Perry, C.J. van Oss and E. Grushka (eds), *Separation and Purification Methods.* Marcel Dekker, New York. Vol. 2, pp. 199-299.

Yarmush, M.L., Antonsen, K. and Yarmush, D.M. 1985. In M. Moo-Yang, C.L. Cooney and A.E. Humprey (eds), *Comprehensive Biotechnology.* Pergamon Press. Vol. pp. 489-506.

24

SAFETY IN MICROBIOLOGY AND BIOTECHNOLOGY

P. HOFFMANN
DSM— German Collection of Micro-organisms and Cell Cultures, Braunschweig, Germany

Introduction

The relations between micro-organisms and man vary from harmless or useful co-existence to infection by dangerous pathogens which may cause severe disease or other problems. The increasing use of micro-organisms in laboratories and industrial production has called for special precautions and legal guidelines to maintain occupational health and product quality when utilizing potentially dangerous organisms. Biological safety is part of a general scheme of quality assessment and control as published by the Organisation for Economic Co-operation and Development (OECD) for micro-organisms (OECD, 1982) and genetically engineered organisms (OECD, 1986) on the basis of numerous publications and expert reports.

RISK ASSESSMENT

Pathogenicity and virulence determine the potential hazard of a micro-organism towards man and his environment. For practical purposes, the natural continuum from harmless organisms like lactobacilli or beer yeast to highly dangerous species like the variola virus has been divided into four hazard groups (Center for Disease Control, 1972) . The group definitions, which may seem to be rather vague, are as follows:

Group 1: No risk. Well known biological agent. Not reported as cause of illness in healthy people or animals.

Group 2: Low risk. Agents that may cause disease in healthy people and animals; regularly present in the environment. Effective prophylaxis and treatment of disease is available.

Group 3: Medium risk. Agents that may cause severe human or animal disease; hazardous if spread into the environment. Prophylaxis and treatment mostly available.

Group 4: High risk. Agents that will cause severe disease. No effective prophylaxis and treatment. Contact and spread must be strictly avoided. So far, only viruses have been classified in this group.

The delimitations of the four hazard groups are not very sharp and there is some dispute about the disposition of several species on the borderline. This is the case especially with organisms that have occasionally been reported in connection with a disease. They may be dangerous only to predisposed people suffering from other severe illness (diabetes, tumour) or undergoing immunosuppressive therapy. These species are present everywhere in nature and, for healthy adults, they are no hazard. From this viewpoint they may be classified as harmless (group 1). If their role as agents of opportunistic infections is stressed, however, they could also be classified as belonging to hazard group 2.

Several characteristics have to be considered to assign a certain species to one of these hazard groups:

1. Host range and susceptibility. Is the organism able to overcome man's immune system? What is the action in immune system? What is the action in immunocompromised hosts? Is a successful vaccination possible?
2. Infective dose. What number of living cells must be incorporated to cause a serious effect?
3. Way of transmission. Is the agent airborne (spores, capsules, etc.), is direct contact necessary, or is it carried by some (animal) vector?

4. Route of infection. Only very rarely is the intact skin the port of entry. Mostly inhalation, ingestion or inoculation have been traced as the route of infection.
5. Kind of disease. In the case of severe illness, what preventive surveillance or curative treatment is available? The production of toxic metabolites or allergenic effects may influence the disposition.
6. Natural occurrence and distribution. Organisms, not present everywhere in the environment must be kept under stringent containment to prevent their spreading into new areas where the immune status of inhabitants may be low or new hosts of economic importance are threatened.

In genetically engineered micro-organisms (GOM), the risk assessment has to include the donor, the host and the vector system. The treatment of organisms that live in or on man (e.g., *E. coli*) is different from other organisms as the impact of incidental contact may be much higher. Deliberate release of GOM is under strong dispute and has led to numerous regulations and guidelines, which need to be harmonized as is now attempted by the various directives of the Commission of the European Community (Leroy, 1991), which will define a common minimal standard for work with natural and genetically engineered organisms.

SAFETY CONSIDERATIONS

The ultimate aim of all actions is to minimize or exclude contact between humans and environment with the organism under study. Depending on the hazard group, four levels of containment have been defined for this purpose. They are summarized in Table 24.1.

The basic level is defined by the 'good microbiological technique' (GMT; Table 24.2), a list of activities which should be self-evident for every laboratory worker. A similar list for the production is the 'good industrial large scale practice' (GILSP; OECD, 1986). Of known hazard are a number of activities which may cause the development of aerosols or the escape of organisms into the working area or the environment. This includes the handling of liquid cultures by mixing, pipetting, etc., the breakage of containers, the opening of culture vessels, the handling of contaminated laboratory animals, unnecessary vibration of inoculation instruments, the preservation of cultures by drying, etc.

TABLE: 24.1
SUMMARY OF RECOMMENDED BIOSAFETY LEVELS.

Biosafety Level	*Practices and Techniques*	*Safety Equipment*	*Facilities*
1	Standard microbiological practices	Note: primary containment provided by adherence to standard laboratory practices during open bench operations.	Basic
2	Level 1 practices plus: Laboratory coats: decontamination of all infectious wastes; limited access; protective gloves and biohazard warning signs as indicated	Partial containment equipment (i.e., Class I or II Biological Safety Cabinets) used to conduct mechanical and manipulative procedures that have high aerosol potential that may increase the risk of exposure to personnel.	Basic
3	Level 2 practices plus: Special laboratory clothing; controlled access.	Partial containment equipment used for all manipulations of infectious material.	Containment
4	Level 3 practices plus: entrance through change room where street clothing is removed and laboratory clothing is put on; shower on exit; all wastes are decontaminated on exit from the facility.	Maximum containment equipment (i.e., Class III biological safety cabinet or partial containment equipment in combination with full-body, air-supplied, positive-pressure personel suit) used for all procedures and activities.	Maximum Containment

Laboratory equipment, protective and building facilities increase in complexity and cost when working with organisms of higher hazard groups. A comparative description is given in Table 24.3.

Legal Aspects

Since the early 1970s, biotechnology has increasingly become the topic of public debate. The fear of creating a 'monster bug' and its potential environmental consequences has led to legislative and regulative action by many governments that also cater for the commercial fate of biotechnology-derived products.

TABLE: 24.2
SUMMARY OF GOOD MICROBIOLOGICAL TECHNIQUES.

- Windows and doors to be closed during work
- Protective clothing should only be worn in the laboratory
- Activities that may generate aerosols (inoculation, filling, blending, mixing, etc.), should be carried out in a biosafety cabinet
- No mouth pipetting
- Avoid needles, if possible
- Regular disinfection of hands, working tables, floors especially in case of spillage or accident
- Eating, drinking, smoking, application of cosmetics, storage of food, and similar activities that may lead to direct contact with biological agents in the laboratory is strictly prohibited
- Contaminated or toxic waste should be collected in special containers and disposed correctly (sterilization, incineration, etc.)
- Proper function of safety equipment must be checked and documented regularly; the laboratory should be kept tidy
- Training of personnel in microbiological techniques; information about potential hazards; emergency scheme in case of accidents
- Access to laboratory for authorized persons only. Hazardous area should be marked by appropriate symbols
- Identity of micro-organisms should be checked regularly

A compilation of relevant data in 12 major countries up to 1987 is given in (Gibbs *et al.*, 1987). As mentioned above, the European Community has decided to set up directives to define a common legal status within its member states until 1993.

Many other domains of legislation have considerable impact on biotechnology: laws concerning waste disposal, handling of toxic substances, environmental protection, occupational health, quarantine regulations, transportation, etc., have to be considered when working with biological agents.

National or interstate shipment of cultures, for instance, are governed by postal and freight regulations, which may be contradictory in some aspects.

A guide to these regulations and the various schemes for packaging and transport of cultures is to be compiled by the Information Centre for European Culture Collections (ICECC, 1991).

As an example, the correct packaging and labelling of agar-slant cultures is given in Fig. 24.1.

TABLE: 24.3
CHECK LIST FOR LABORATORY SAFETY LEVELS. SYMBOL DASH (-) MEANS NO SPECIAL MEASURES NECESSARY.

Description	*Safety provisions*			
	L1	*L2*	*L3*	*L4*
Physical separation	in some cases	yes	yes	separate building wherever possible
Closed system		yes	yes	yes
Special control area	- -	in some	yes cases	
Hazard warning sign	-	in some cases	yes	yes
Windows and doors	no special measures	kept closed while work is in progress	closed	sealed unbreakable glass
Ventilation to prevent contamination of the outside air	no special measures	no special measures	in some cases	yes, HEPA filters
Negative pressure in the rooms	-	- -	>30Pa	30-50 Pa, graduated
Intake and exhaust-air via filters	-		yes	yes, double system
Access restrictions	-	permission required	limited; via airlock system	employees only; via airlock system
Facilities for personal washing and decontamination	wash-basin accessible	wash-basin in lab. + disinfection	airlock + shower wash-basin + disinfect	airlock+shower + disinfection wash-basin in each room
Shower before leaving	-	-	in some cases	yes
Collection and inactivation of waste water	-	-	in some cases	yes
	-	-	in some cases	yes
Protective clothing	in some cases	lab. gown	yes, change of clothing	full- protection
Biological safety cabinet	in some cases	Class I or I	Class II or II	suit Class III
Sterilization capacity	accessible	in the bldng.	in the lab.	double-door sterilizer in the lab.
Disposal of	in some	inactivation	inactivation	kill tank,

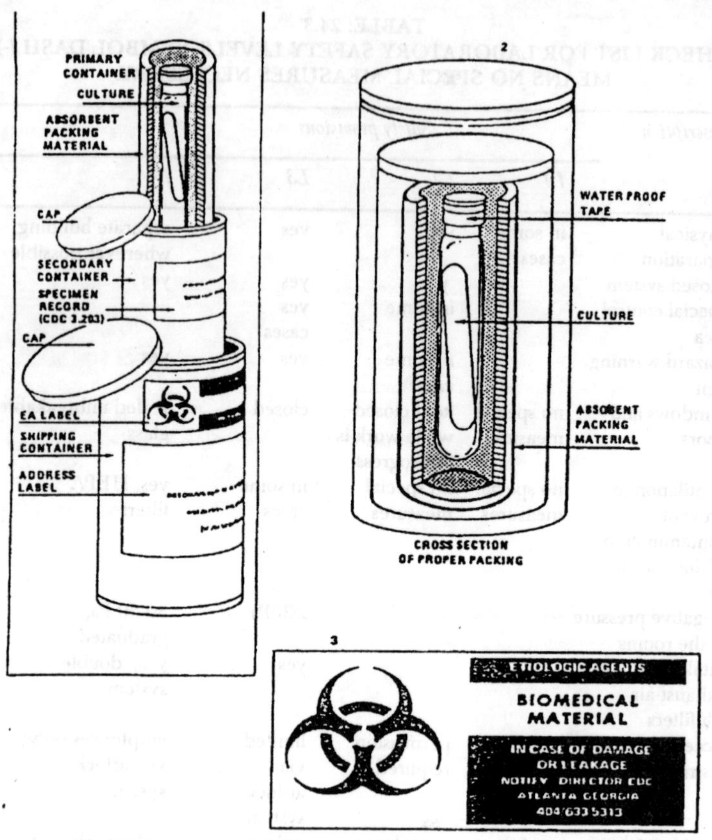

Fig. 24.1: Packaging scheme for agar cultures of micro-organisms. Exploded view (1), cross- section (2) , and biohazard (3).

CONCLUSIONS

Experience has shown that it is possible to work with biological agents without exposing employees and the environment to hazards as long as all safety guidelines are respected. Regular staff training and the motivation of everyone involved in this field to follow the guidelines and regulations is of paramount importance for safety in microbiology and biotechnology.

REFERENCES

Gibbs, J.N., Cooper, I.P. and Mackler, B.F. 1987. *Biotechnology & The Environment: International Regulation.* New York.

Information Centre for European Culture Collections (ICECC). 1991. *Instructions for the Shipping of Infectious and Non-Infectious Biological Substances.* Braunschweig, Germany.

Leroy, O. 1991. *Biotechnology: EEC Policy on the Eve of 1993.* European Study Service, Rixensart.

OECD. 1982. *Guidelines for Testing of Chemicals.* OECD, Paris.

OECD. 1986. *Recombinant DNA Safety Considerations.* OECD Publicat. No. 43691. Paris.

U.S. Department of Health, Education and Welfare: Classification of Etiologic Agents on the Basis of Hazard. 1972. Center for Disease Control, Atlanta, Georgia 30333, USA.

25

THE SOUTH-SOUTH DIMENSION IN PARTNERING: STRATEGIES ALLIANCES IN THE BIOTECHNOLOGY SECTOR

CARLOS M. CORREA

Director, Science and Technology Management Graduate Study PRogramme University of Buenos Aires Buenos Aires, Argentina

1. INTRODUCTION

Despite the achievements in the application of biotechnology to agricultre and other fields, the pharmaceutical industry is the area where the commercial use of biotechnology still is the most significant. The importance of biotechnology to the pharmaceutical industry is twofold.* First as a production technology, it has led to the introduction of a number of products now on the market, and many others which are in the pipeline waiting for approval by health authoriteis. Second, and perhaps

---

*This article is based on research conducted by the author (see Correa, 1991 and 1994) and on the results of a research project on firm strategies and public in biotechnology in Latin America" carried out at the University of Buenos Aires with the support of the International Development Research Centre (IDRC) of Canada.

more importantly, biotechnology haas modified, as a research method, the research paradigm of the pharmaceutical industry.

Conventional biotechnological products account for about 30 per cent of the world pharmaceutical market. Antibiotic have a major share, followed by steroids, vitamins and vaccines. Modern biotechnology-based pharmaceutical products now on sale included about 20 molecules, without considering diagnostics kits, which are an important segment of biotechnology-based products (Thomas, 1993, p. 5; Diaz, 1997). Erythropoietin has become the top selling biopharmaceutical, followed by the vaccine against hepatitis B, recombinant insulin, human growth hormone, alpha interferon, granulocyte conoly-stimulating factor (G-CSF) and tissue plasminogen activator (TPA). These products account for a major share of the total pharmaceutical biotechnological market.

In global terms, the impact of modern biotechnology on the pharmaceutical market is still modest; biopharmaceuticals account for no more than 2 to 3 per cent of the world market. This does not reflect, however, the importance of modern biotechnology for the pharmaceutical industry. It has already had a major impact on the way new drugs are designed and developed.

The pharmaceutical industry is one of the most R & D- intensive sectors in the world economy. The development of new compounds and the discovery of new therapeutical uses of old drugs is esential in the competitive strategies of major firms. Units the 1970s, mass screening of potentially useful molecules had enabled a high rate of innovation to be maintained. Mass screening, however, is long and costly, and highly aleatory. The number of new drugs that show a clear therapeutical advance startecl to tall in 1975 (Teitelman. 1989, p. 29). Biotechnology opened a new research paradigm: the "rational drug design". Instead of testing thousands of molecules until a useful one is identitied. the new approach makes it possible to target the biochemical reactions that a disease triggers, and then to devise a chemical compound to stop them. Rational drug design is based on the understanding of the physiological basis of a disease. Research focuses on the activity, of enzymes, hormones and hormone receptors, cell replication and protein synthesis, and other molecular-level aspects ot disease and drug treatment (OTA, 1991, p. 74). One of the main new techniques of research protein engineering makes it possible to replace, insert or delete amino acides in naturally occurring amino acids, manipulate, domain structures of proteins or design proteins *abinito.*

The outcome ot rational drug design, though based in biotechnology, is the production of synthetic molecules that are able to interact in the disease process. Rational drug design not with standing its potential, is not likely to completely displace traditional screening methods; rather, it will complement them (Blundell, 1994, p. 148). The development of "combinatorial.) chemistry may substantially reduce R&D time and costs in drug discovery. With this new technology it is possible to create simultaneously compounds with shapes, charges and electrostate characteristics covering an enormous range of possibilities, and assay the resulting molecules all at once to identify that look promising for a particular, predetermined aim (Alper, 1994, p. 399).

The entry by pharmaceutical firms into biotechnology R&D has combined a variety of approaches. They include: (i) establishing their own R&D centres, as in the case of Merek, Schering Plough, ICI, Sandoz and Ciba Geigy (now Novartis); (ii) acquiring specializeid biotechnology firms: (iii) establishing research contracts with laboratories at universities; (iv) setting up other forms of strategic alliances. The adoption of biotechnology by the pharmaceutical industry has been helped by a massive effort in R & D, both private and public. (v) entering into licensing or marketing arrangements with specialized biotechnology firms; and (vi) setting up other forms of strategic alliances. The adoption of biotechnology by the pharmaceutical industry has been helped by a massive effort in R & D, both private and public. Worldwide, the pharmaceutical industry spends more than US$ 20 billion annually on R & D. The United States Federal Government alone allocated US$ 1,742 million to health research in the framework of he Federal Biotechnology Research Initiative of 1994.

II. CHALLENGING THE DOMINANCE OF THE PHARMACEUTICAL GIANTS

Biotechnology has been regarded as an opportunity for new firms to enter into the pharmaceutical industry. To what extent has biotechnology actually created a challenge to the position of the established oligopoly in the pharmaceutical industry? The potential application of biotechnology to a vast number of areas gave rise, during the second half of the 1970s, to the emergence of biotechnology "start-ups" (i.e. specialized biotechnology firms), particularly in the United States. Many of them were established and staffed by reputed scientists. However, the promises of biotechnology did not materialize as expected. The emrging firms soon faced financial constraints and the anxiety of investors for a return in the short term (Correa, 1989, p. 987). After an initial euphoria, which allowed many of those firms to obtain public funding, a process of take-overs and acquisitions gave large firms from various sectors control over the start-ups. Even the most successful companies, in terms of technological developments, had to license their new products to established firms, or to accept equity participation to survive. One example is the case of Genentech, one of the most innovative start-ups, which eventually evolved into a fully integrated pharmaceutical (or, more accurately, biopharmaceutical) company. It built up a sales network in the United States from scratch, and started to do the same in Europe, in order to commercialize its "star product", the TPA clot-buster "Activase". These efforts were however. insufficient; Hoffmann La Roche acquired a majority share in Genentech.

Thus, the challenge from small, technologically strong companies did not disturb the dominance of the large firms, which rapidly obtained, through various channels, direct or indirect control over many of the new and fragile firms. Many small biotechnology firms, however, have developed complementary links with large pharmaceutical firms, particularly through strategic alliances. Such alliances offer a way for the small companies to survive and develop, and offer the large companies a flexible way to farm out research on biotechnology.

In sum, notwithstanding the disturbing potential of the new technology - in terms of Schumpeterian analysis - the solid established pharmaceutical oligopoly rapidly reacted to maintain, and even strengthen, its market position. This strongly suggests that technological skills if available, do not in themselves guarantee in a successful economic and commercial operation in biotechnology. As in other areas of new technology, distribution and marketing can be far more difficult barriers to surmount than barriers of a technical nature. The amount of resources needed and the expense of R&D can also be serious constraints for new potential entrants.

Can biotechnology create or enhance the chances of developing countries successfully entering the pharmaceutical market? Several characteristics of biotechnology suggest the answer to this question may be yes. First, capital requirements for production are relatively low in comparison with those for other industrial activities, particularly in the field of other new technologies such as microelectronics. The investment necessary for infrastructure and equipment for a small plant may be in the order of US$ I million. Of course, capital requirements may increase as the industry matures, but then economies of scale in production and other advantages of size also increase.

Second, the techniques of genetic engineering have become routine and the availability of skilled personnel may not be problematic in many developing countries where a good base exists in biosciences. If the original development of a biotechnology-based drug is sought, however, the barriers of entry will perhaps be too high for most developing countries. If an imitative path is followed instead, the entry costs would be affordable. One Argentine firm, for instance, was able to undertake the production of recombinant interferon, based on access to published scientific information and a well-qualified staff in contact with foreign research institutions. Given that the genetic sequence of interferon was already known and published, that firm developed a method of isolation, without entering into the costly and lengthy screening process that Biogen originally had to undertake.

Third, entry by firms in developing countries is facilitated by their access to biological materials (cloned genes, cloning and expression vectors, recombinant viruses, etc.), which are available in the market at reasonable cost. In addition, modern electronic equipments now makes it possible to undertake more economically and rapidly tasks that in the past were lengthy and costly (such as the synthesis of nucleotides). Some specialized firms are even offering kits with specified properties to clone a desired gene or are undertaking gene-cloning on request, for a small fee. In other words, the diffusion of knowledge and the market availability of biological inputs and specialized services and equipment have considerably facilitated gene-cloning and thereby reduced the entry barriers.

There are, however, many important constraints. First, though laboratory techniques can be mastered, scaling up may pose difficulties insofar as industrial microbiologists and other specialists in the downstream activities are scarce or non-existent. Second, the general infrastructure available in the country where the firm operates also plays an important role. For example, since most enzyme and related materials are unstable at ordinary temperatures, facilities for cold transportation and storage are essential, not to mention more basic requirements regarding water and electricity supply. As in other undertakings, the overall industrial infrastructure for equipment repair, recruitment of technical staff for laboratories and so on are equally important to ensure efficient production. Third, the marketing of biotcehnology-based products may pose serious barriers for firms which do not possess a sales network and the necessary financial resources. (Fransman, 1989, p. 10). Fourth, the commercialization of pharmaceutical products requires the approval by national health authorities in each country where the product is to be sold. Obtaining such approval is a complex task in many jurisdictions, particularly in those where legislation grants the first firm to register a product exclusive rights to the use of test data developed with respect to such product.

Another important barrier is now emerging in developing countries as a result of the expansion as the strengthening of intellectual property rights. In response to pressure from the United States (under section 301 of the United States Trade Act) or in order to conform to the standards adopted under the Agreement on Trade-Related Aspects of Intellectual Property Rights (the "TRLPs Agreement") negotiated during the Uruguay Round, developing countries are introducing patents on pharmaceutical products. They must also protect micro-organisms, which are generally understood in patent law as comprising cells and subcellular components alike.

These changes will prevent or make more difficult imitative path of development, whenever pattern are granted with respect to products. 'They will also enhance the market power of large pharmaceutical firms to determine the conditions for voluntary licences, unless an effective system of compulsory licences is established. Several laws recently enacted in Latin America (e.g. in Argentina, Chile and the countries of the Andean Group) provide for different types of compulsory licences.

The extension of patent protection to biotechnology-based products, as granted in industrialized countries, may create additional difficulties for developing countries.

Genetic engineering now makes it possible to produce recombinant products identical to those existing in nature. Their patentability per se is, hence, questionable. An exarnple of the kind of conflict that may arise is provided by the case of Genetech's patent on TPA, which was revoked by the United Kingdom High Court in litigation with Wellcome, but deemed valid in the United States. This case showed clear differences in the judicial interpretation of what is patentable in the area of recombinant technology. Another, even more striking case, was the revocation in October 1994 by the United Kingdom Court of Appeal of a Biogen patent relating to the hepatitis B vaccine, whereas the same patent was deemed valid in July 1994 by the Technical Board of Appeal of the European Patent Office (Reid, 1995, p. 98).

On the other hand, patent claims are often defined in such broad terms (the Cohen-Boyer patent for E. Coli is a classical example) that patent conflicts are inevitable. The breadth of claims is one of the main sources of litigation.

Conflicts may also arise between patents covering a drug produced by biotechnology and the same drug produced by conventional means. In Scripps Clinic and Research Foundation vs. Genentech, for example, a United States court ruled that Genentech's recombinant Factor VIII (a blood product commonly used to treat haemophilia) infringed the patent granted to the claimant on the basis of a method relying on the purifying of blood. The court, however, allowed Genentech to continue work on the product on the basis that the public interest would be better served if companies other than the patent-holder could continue research on it (Robbins-Roth, 1987, p. 19).

Developing countries are becoming involved in these battles, but have very little to gain therefrom. In several Latin American countries, for instance, administrative and judicial procedures have been initiated by the Genetics Institute against local companies in respect of the production and commercialization of erythropoietin. The Genetics Institute was granted several patents in these countries relating to erythropoietin, though *its* rival in this field, Amgen, had prevailed in the United States and Europe over the Genetics Institute in a series of patent battles based on Amgen's priority in sequencing the respective gene (Cardenas y Espinosa, 1997, p. 56).

III. OPPORTUNITIES FOR DEVELOPING COUNTRIES

As discussed above, a firm in a developing country faces a number of barriers to enter in the field of biotechnology for pharmaceuticals. Original R&D, marketing and distribution generally constitute major constraints. Large established firms can exploit significant economies of scale in these areas. Even with conventional technologies, these factors have posed insurmountable obstacles to firms in developing countries and very few countries (e.g. Argentina, Egypt, India) have been able to develop local pharmaceutical industries of a reasonable size, based on the non patentability of pharmaceutical products. Marketing power and their innovative capabilities have given the large firms market positions which are hard to challenge. According to some analysts only a few big firms will be able to make a profit and consolidate their position in biotechnological pharmaceuticals in the long run (Wemer, 1987, p. 1,093; Schutz and Eichin, 1989, p. 55).

However, a few examples (such as Argentina and Cuba) show that the application of biotechnology to pharmaceuticals is an opportunity that developing countries should not miss. Depending on the type of strategy followed, investments can be affordable and, at least in some countries, good scientific interdisciplinary teams may be built up, including personnel specialized in immunology protein chemistry, biophysics, industrial microbiology and computer sciences. The availability of scientific personnel, nowever, may not necessarily be the critical point. Scaling-up i.e. the effective implementation of bioreactor and bioprocessing technologies, requires industrial skills that are not easy to acquire. Scaling-up is also one of the more expensive and time consuming phases of biotechnological production and may constitute a major barrier to entry.

While larger companies can make clear economies of scale in R&D activities, production can be efficiently undertaken on a small scale for many low-volume, high value-added products. Flexible multipurpose plants can produce a wide range of products without the need for additional investments each time a new substance is to be produced. In addition, once a biocapability has been established, it is possible to take advantage of the substantial "economies of scope" that biotechnology offers, by shifting from pharmaceuticals to diagnostic kits, food processing, veterinary and other products.

Access to foreign technology may not be easy, since technologically innovative firms are wary of creating potential competitors through licensing (Correa, 1994). Innovative start-ups, however, generally need to license the technologies they have developed in order to generate revenue and continue financing R&D. They are, therefore, a potential source of technology. However, due to the knowledge-intensive nature of biotechnology and the important role played by university research, it is common for several enterprises and university laboratories to undertake R&D on the same subject at the same time. Scientific publications and formal or informal contacts with researchers at those laboratories may generate valuable information and technology.

Finally, finding a role to play in biopharmaceuticals requires the development not only of scientific and technological skills but also crucially, of marketing capabilities, including through joint ventures and other from of cooperation. Many firms are able to develop good products; their main problem is how to sell them on the international market.

IV. THE ROLE OF STRATEGIES

There is all extensive literature on the important role that different types of strategic alliances have played in high-technology industries. Biotechnology seems to be one of the areas-second only to information technology - where strategic alliances have been most frequent (UNC5AD. 1998). Several factors explain the important role of straiegic alliances in the field of biotechnology. One of the major reasons relates to capital needs. However, the most prominent factor seems to be related to the exploitation of marketing capabilities (Burrill and Lee, 1990, pp. 42-43). Quite often the institution or firm that has developed a certain technology lacks the resources and skills necessary to market the resulting product. This has generally been the case of the "start-ups" mentioned above: With a few exceptions, most of those firms were not able to develop their own marketing mechanisms, especially given the large investments required to constitute and maintain a sales force in the pharmaceutical field. An additional problem is that the marketing of a new pharmaceutical product requires complimance with complex registration procedures. This is an area in which the contribution of an experienced partner may be essential, particularly when dealing with national authorities in countries with high health standards.

Another typical form of strategic alliance relates to R&D activities. These alliances often take place between rivals in the market who find cooperation rather than competition a better way to undertake large or risky technological projects. Such alliances allow partners not only to share costs, but to complement each other's skills and speed up the R&D process.

Marketing-oriented alliances (those based on R&D as well as other types of alliances) may assume "vertical" and "horizontal" modalities, depending on the market position o each partner and its contribution to the joint operation (Fernandez Sanchez, 1996, pp. 351-353). Vertical alliances are based on specialization: for instance, one of the partners undertakes research and production and the other the marketing of products in agreed markets. "Horizontal" alliances involve two types of arrangements: competitive horizontal alliances" in which partners are direct competitors in the market, but decide to collaborate rather than compete in a specific area; and (ii) "complementary horizontal alliances" in which each of the partners has assets that are complementary to those of the other.

As indicated in the vast literature on the subject (see UNCTAD, 1996), most strategic alliances are between firms and other institutions in developed countries. There are many good reasons why South-South cooperation in the form of strategic alliances may be appropriate and beneficial for the partners and countries involved. They include the existence of similar technological levels, similar diseases, like availability of relatively low-cost personnel, and a more balanced bargaining position between partners than in cases where firms from the North are involved.

The agreements entered into between an Argentine pharmaceutical company (ELEASE) and two Cuban R&D institutions, which are briefly reviewed below, provide an example that strategic alliances may play in South-South relations. Though it is too early to assess the results, of this experience, it certainly indicates that the flexible business framework offered by strategic alliances may, under certain circumstances, benefit firms developing countries.

V. EXAMPLE OF A SOUTH-SOUTH PARTNERSHIP

A. The parties

Cuba has, since the 1958 revolution, emphasized the development of its scientific and technological capacity. Despite the small size of the economy and the impact of the economic blockade, Cuba has the highest rate of R&D personnel per million inhabitants in the world, and the country's R&D expenditure in relation to the GNP is one of the highest in Latin America and the Caribbean (UNESCO, 1991). The Cuban Govermnent has identified work on biotechnology as a national priority. In 1986, it established a centre for genetic engineering and biotechnology, the Centro de Ingenieria Genetica y Biotecnologia (CIGB), to work in different areas of biotechnology research and production, including biopharmaceuticals, diagnostic kits and plant biotechnology. With a modern infrastructure, CTGB operates its own plant with a staff of over 700 people. Production costs, particularly in terms of the salaries for research and technical personnel, are significantly lower than in other Latin American countries, and far lower than those in developed countries. CIGB has developed recombinant technologies for the production of interferon, streptokinase, hepatitis B vaccine, epidermic growth factor and several diagnostic kits, among other products. Heber Biotec SA was founded in 1991 as a specialized firm to market CIGB products domestically and internationally.

In addition to CIGB, at least three other institutions undertake research and production in the biotechnology area. The Centro de Inmunologia Molecular (CIM)– Centre for Molecular Immunology- has undertaken substantial work in the field of monoclonal antibodies. The Centro Nacional de Biopreparados (National Centre for Biopharmaceuticals) has developed vaccines and other biopharmaceuticals. The Finlay Institute has also developed vaccines for the prevention of several diseases, such as meaningitis. The scientific and technical achievements of these institutions are widely recognized in the region (Diaz, 1996). However, the marketing of their products has faced several of the problems mentioned above, aggravated by the economic blockade enforced by the United States Government. Market access has been one of the most crucial issues, particularly in the biopharmaceuticals field. In order to sell any product, national health regulations need to be complied with. In the case of vaccines, they are generally acquired by government agencies on the basis of bidding procedures. Cuban producers faced the mistrust that is often associated with production originating in developing countries.

The Argentine market for pharmaceuticals amounted to around US$ 5.5 billion in 1997. It is the 12th largest pharmaceutical market (by value of sales) in the world. ELEA SA is a locally owned Argentine pharmaceutical company with annual sales of over US$ 40 million. It markets a wide range of products, including several products to combat the herpes virus and has connections with severals university research centres.

B. The Alliances

CIGB, CIM and the Finlay Inistitute had developed a strong R&D and industrial capability, but faced enormous bariers to market their products. ELEA SA realized the advantages that an alliance with those institutions may bring about. The most urgent need of the Cuban institutions was to find commercial channels for their products, while ELEA would be able to expand its mix of products and consolidate its market position in Argentina.*

Several arrangements have been entered into between ELEA and the Cuban institutions in the last five years. First, the Finally institute agreed to produce the anti-meningococcical vaccine for distribution and sale in 'Argentina through ELEA. ELEA and the Institute agreed to conduct field tests of the vaccine in-Argentina, following the standards of the Center for Disease Control of the United States. ELEAS' sales' of the vaccine against meningitis produced in Cuba reached US$ 1.1 million in 1997. Though this agreement was centered around the marketing by one partner of product developed and produced by the other, the technical capabilities and financial support of the Argentine partner allowed the Cuban partner to undertake tests on the vaccine that are likely to strengthen its market prospects in Argentina and elsewhere.

Second, CIGB agreed to market various biopharmaceuticals it produced, such as interferon and the hepatitis B vaccine, in Argentina exclusively through ELEA. This arrangement is a typical case of complementarity of production and marketing capabilities.

Third, and most interestingly, ELEA and CIM entered into a research agreement to develop and market an anti-cancer product, based on technology originally developed by CIM. Based on a network of Argentine research institutions and specialists, a project was set up in Argentina in order to undertake the different phases required for the finalization and approval by the health authorities of a new product.* In 1998, clinical tests on humans were due to start. In this project the partners were able to complement their abilities in different fields. While Cubans contributed the design of the new drug, the Argentine company was able to put together the team of

---

* It should be noted that pharmaceutical product patents will be enforceable in Argentina after the year 2000. This will significantly change the patterns of competition in the market and, in particular, limit the access by local companies to new drugs, which will generally be subject to the patent rights of the innovative firms.

researchers needed to test the product. Since these two activities (design and testing) require different resources and skills, the partners found ways to complement each other.

In sum, though the commercial aspect seem to have been dominant at the initial stage of the relationship between the partners, it has gradually evolved towards a more complex arrangement in which partners also exploit their complementarity in R&D capabilities.

VI. CONCLUSIONS

The impact of biotechnology is already pharmaceutical field, in terms of new products and, perhaps most importantly, the radical change in the industry's research paradigm. Despite expectations in some developing countries, biotechnology, as applied to pharmaceuticals, has not meant a challenge to the dominant position of major firms. Economies of scale in research and marketing, among other factors, countinue to create high barriers to entry. Moreover, the recent wave of takeovers and fusions points to an increasing concentration of market power.

Developing countries willing to use biotechnology as a platform for creating new activities or expanding the existing ones face many difficulties (not necessarily of a technical nature) including the implications of the current patterns in intellectual property protection. Strategic alliances, especially those that exploit the complementarities between partners, may offer the means to face some of the constraints. The case presented above illustrates how such alliances may permit relatively small companies in the South to apply biotechnology commercially in the field of pharmaceuticals, including through undertaking original R&D.

REFERENCES

Alper, J. (1994). "Drug Discovery on the Assembly Line",; Science, Vol. 264, (June).

Blundell, Tom L. (1994). "Problems and solutions in protein engineering-towards rational design", *TIBTECH,* Vol.12, (May).

Burril, G. and Lee, K (1990). *Biotech* 91: *A Changing Environment. Fifth Annual Survey of Business and Financial Issues in America's Most Promising Industry.* San Francisco: Ernst & Young.

---

* The project has been entirely financed by the partners. The local firm has applied for an authorization to use a tax credit system, recently announced by the Argentine Government to promote local R & D.

Cardenas y Espinosa, R. (1997). ***Aspectos Tecnologicos de las Patentes.*** Mexico, D.F.

Correa, C. (1989). "Biotecnologia: el surgimiento de la industria y el control de la innovacion", ***Comercio Exterior, Vol.*** 39, No. I 1, Mexico, D.F.

Carrea, C. (1991). "The pharmaceutical industry and biotechnology: opportunities and constraints for developing countries", ***World Competinon,*** Vol. 15, No. 2.

Correa, C, (1994). "Trends in technology transfer: implications for developing countries", ***Science and Public Policy,*** Vol. 21, No. 6.

Diaz, A. (1996). "Biotecnologia en la Industria Farmaceutica de America Latina: Panorama, Realizaciones y Perspectivas", ***Jornadas Interamericanas de Ciencias Farmaceuticas.***

Diaz, A. (1997). "Biotecnologia industria. Una Alianza Estrategica", ***in-Encrucijadas.*** Buenos Aires: UBA.

Fernandez Sanchez, E., (1996). ***Innovacion, Tecnologia y-Alianzas Estrategicas. Factores Clave de la Competencia.*** Madrid: Biblioteca Civitas Economia y Empresa.

Fransman, M. (1989). ***Industrial applications of biotechnology. Implications for industrial policy in developing countries.*** Vienna: UNIDO.

OTA (Office of Technology Assessment) (1991). ***Biotechnology in a Global Economy.*** Washington, D.C.

Reid, Brian C. (1995). Biogen in the EPO. The Advantage of Scientific Understanding, ***European Intellectual Property Rights*** (EIPR), No.2.

Robbins-Roth, C. (1987). "Patents vs. public interest", ***High-technology Business,*** (December).

Schutz, S. and Eichin, K (1989) "The significance of genetic engineering for the pharmaceutical industry", ***Drugs Made in Germany,*** Vol.32, No. 2.

Teitelman, R. (1989) "Pharmaceuticals. Global Report", ***Financial World, (30*** May).

Thomas, Sandra M. (1993). ***Global Perspective 2010. The Case of Biotechnology.*** Brussels: Fast.

UNCTAD (1996). ***Science and technology issues. Emerging forms of technological cooperation: the case for technology partnership.*** Geneva and New York: United Nations Publications, UNCTAD/DST/13, United Nations Sales No. E.96.II.D.1.

UNCTAD (1998). ***World Investment Report 1998. Trends and Determinants.*** Geneva and New York: United Nations Publications, United Nations Sales No. E.98.II.D.5.

UNESCO (1991). ***World Science Report,*** Paris: UNESCO.

Werner, R. (1987). "Biobusiness in the pharmaceutical industry". ***Drug Research,*** Vol 37 (II), No. 9.

Source: United Nations conference on Trade and Development, at as XI Bulletin, New Approaches to Science and Technology Cooperation and Capacity Building, Document No. UNCTAD/ITE/EDS/6, United Nations, New York and Geneva, 1990, pp. 289-299.

•••